UNSEEN COSMOS

UNSEEN COSMOS

the universe in radio

FRANCIS GRAHAM-SMITH

OXFORD
UNIVERSITY PRESS

Great Clarendon Street, Oxford, OX2 6DP,
United Kingdom

Oxford University Press is a department of the University of Oxford.
It furthers the University's objective of excellence in research, scholarship,
and education by publishing worldwide. Oxford is a registered trade mark of
Oxford University Press in the UK and in certain other countries

First Edition published in 2013

Impression: 1

Published in the United States of America by Oxford University Press
198 Madison Avenue, New York, NY 10016, United States of America

British Library Cataloguing in Publication Data
Data available

Library of Congress Control Number: 2013940786

ISBN 978-0-19-966058-2

Printed in Great Britain by
Clays Ltd, St Ives plc

Contents

Preface

Radio telescopes have transformed our understanding of the Universe. Pulsars, quasars, Big Bang cosmology, all are discoveries of the new science of radio astronomy. My life has been devoted to this subject since my first days as a research student, in 1946, and I have seen radio astronomy grow from the first discovery of cosmic radio waves to its present role as a major part of modern astronomy. Today a new generation of radio telescopes promises to extend our understanding of the Universe into further, as yet unknown, fields. Huge new radio telescopes are being built, with exotic names such as Atacama Large Millimetre Array (ALMA), Low Frequency Array for Radioastronomy (LOFAR), and the Square Kilometre Array (SKA). Radio telescopes on spacecraft such as the Cosmic Microwave Explorer (COBE) and Planck are tracing in minute detail the faint but universal radio signal from the expanding early Universe.

Radio is part of the electromagnetic spectrum, which covers infrared, visible light, ultraviolet, X-rays, and gamma-rays. In this book I will explain why it is that radio waves give us a unique view of the Universe. I will trace the development of radio telescopes, showing how each new idea in observing technique has led to new discoveries. I will look at the way in which radio waves are generated in the various cosmic sources, and relate these processes to the more familiar radio world of mobile phones, radio and television channels, wireless computer connections, and remote car locks.

International collaboration is a natural part of the radio astronomer's lifestyle. There are close liaisons with colleagues in related disciplines; X-ray and gamma-ray observatories orbiting in space are observing the same objects as ground-based radio astronomers working with radio. There are also direct connections between radio observatories in different countries, constructing networks of telescopes which have capabilities way beyond those of their individual components. The culmination of these ideas, the Square Kilometre Array, cannot be built by one country alone; in fact parts of it will spread across distances of thousands of kilometres. I find these connections and networks an inspiration and an example in international collaboration. There can be few more exciting branches of science to be in at this time. I hope that through the following pages you too will share in the excitement of discovering the wonders of the radio universe, and the possibilities promised by the new age of giant radio telescopes.

In this wide-ranging survey I have been greatly helped by colleagues at Jodrell Bank Observatory, particularly by Tom Muxlow and Ian Morison. Professor Malcolm Longair of the Cavendish Laboratory read a complete draft, correcting errors particularly in cosmology. My wife Elizabeth, who having been assistant to Martin Ryle in 1945–1946 is a true veteran of radio astronomy, has brought lucidity to a complex text. My thanks to all these, and to Latha Menon and Emma Ma and the team at Oxford University Press.

1

Radio Noise from Space

In 1610 Galileo published the first observations of the sky with the newly invented telescope. Before then the heavens were thought of as a pattern of fixed stars imprinted on an inverted hemispherical bowl. The Sun, the Moon, and the planets moved slowly across this surface, following geometrical rules which revealed little of the true nature of the solar system and the stars beyond. It was Galileo's discovery of the moons in orbit around Jupiter that transformed this picture into a fully three-dimensional universe. His revelation of the value of telescopic observations started an evolution of optical telescopes which is still continuing. Astronomy through optical telescopes has progressively revealed our place in the Solar System, the Sun as a star in our local Galaxy, the Galaxy as one among a multitude of extragalactic nebulae, and the evolution of the whole Universe over a cosmological time scale of billions of years.

The new science of radio astronomy arrived 340 years later, in the mid-twentieth century. When I started research in 1946 it was not at all obvious that radio would have any substantial contribution to make to our understanding of the Universe on any scale, except perhaps there might be something to be learnt about the Sun from its outbursts of radio emissions associated with sunspots and flares. There was no conception that radio could contribute to cosmology; although, as we will see, it now provides the most fundamental observations of the early Universe and its evolution. At that time cosmology was very short of observational material. Edwin Hubble had shown in 1929 that the observable Universe was not static; extragalactic nebulae were, on average, receding from us, and receding at a rate which increased with distance. On this one observation, that the Universe was expanding, depended the whole of cosmology. There was plenty of theory, speculating on the past and future of this expansion: how did it start, and would the force of gravity be enough to slow down or even halt the expansion?

It seemed that the further back one looked at the history of the expanding Universe the greater the concentration of matter, suggesting that everything was expanding from a time of infinitely large density. This explosive creation was named the 'Big Bang' by Fred Hoyle, but he challenged this interpretation by speculating that on the largest scale of time and space there might be a 'steady state' of continuous renewal rather than a catastrophic beginning. Radio astronomy has transformed this situation, removed the possibility of a steady state, and provided some precise measurements of the fundamental parameters of the early Universe. We now see cosmology as an exact science rather than a realm of speculation.

In this book I trace the development of radio astronomy from the first observation of cosmic radio waves, through the growing understanding of their origin in many different kinds of astronomical bodies, to the unique contribution made by radio to cosmology. This progression has occurred during my own lifetime, and my account follows my own developing understanding of astronomy, seen as it were through a newly opened window. It is based on personal experience of a development from the simplest early radio observations of the Sun to the multi-element telescopes which now map thousands of remarkable radio sources, such as pulsars and quasars, both within and beyond our Galaxy. The success of radio astronomy has been such that it is no exaggeration to say it has revealed a radio Universe. The importance of the new understanding which this has brought is demonstrated by the scale of the new telescope systems which are being developed by a huge international effort. These are the Low Frequency Array (LOFAR), the Atacama Large Millimetric Array (ALMA), and the Square Kilometre Array (SKA), which will be referred to throughout this book as the culmination of a growing understanding of technique and capability.

Radio is at the long wavelength end of the electromagnetic spectrum, which extends through infrared, visible, and ultraviolet light to X-rays and gamma-rays. Radio, however, is not obviously a useful medium for mapping the Universe; after all, we do not expect to map any part of our own world by detecting the radio waves it emits. Light works well enough: why go to the trouble of using radio, especially when the long wavelengths are such a disadvantage for mapping fine detail? Two answers will emerge in this book. First, radio draws the attention of astronomers to objects which would be impossible or very difficult to discover by optical means, objects such as pulsars and quasars, which are the seat of some of the most energetic processes in the Universe. And, second, because the low frequencies of radio waves lend themselves to sophisticated processes which are not available to the high frequencies inherent in light; radio telescopes have developed far beyond the concepts of conventional optical telescopes. These processes depend on the manipulation of the electric fields of radio waves, and the combination of these fields in extensive arrays of telescopes in ways which are almost impossible in optical telescopes, even at the longer wavelengths of the infrared.

Let us start at the beginning, with the discovery that any detectable radio waves at all are emitted by celestial bodies, and that they can be detected on Earth with simple apparatus.

Jansky's Merry-Go-Round

Long before the satellite links and fibre optics which today carry our international telephone and television, long-distance telephones were using short-wave radio. Sometimes intercontinental radio worked well, but there was always a background noise to contend with. This was regarded as a nuisance to be avoided if at all possible; but first its origin had to be found.

In 1932, when Karl Jansky of the Bell Telephone Laboratories in New Jersey was the first to investigate this background noise, long-distance radio links were using wavelengths of around 15 metres, about the shortest for which technology existed. The radio links suffered both from a continuous background noise and from 'atmospherics'. These are crackling noises from lightning strikes, and the idea was that they might be avoided if directional antennas could be steered away from thunderstorm areas. Jansky built a directional antenna, which he called the merry-go-round, which could be steered round the horizon to locate the origin of atmospherics (Figure 1). It achieved its purpose, but Jansky also noticed that the continuous noisy background radio signal was stronger from some parts of the sky, so that he could use his steerable antenna to find its origin.

Jansky found that the noise, which sounded like a hiss, came from outside the Earth. His steerable antenna, built of brass tubing on a timber framework, was only crudely directional, being only two wavelengths long. By patiently scanning the sky with his rotating antenna, he made rough measurements of the azimuth where the noise was greatest, and he made a crude plot of its strength over the whole 360° of azimuth. The key observation was that this pattern

(a)   (b)

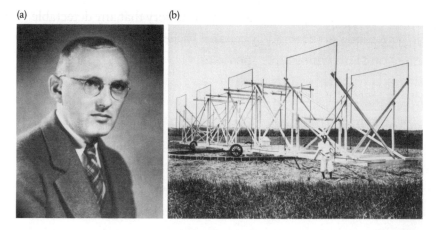

FIGURE 1 (a) Karl Jansky. A photograph in 1928, when he was starting work at the Bell Telephone Laboratories. (b) Jansky's antenna array. The array was 29 metres long, and could be rotated on the circular track. *Image courtesy of NRAO/AUI.*

moved by about 1° every day, slipping by a whole day in the course of a year. This showed that it came from a region of sky that was fixed in relation to the stars and entirely outside the Earth and the solar system. The scans showed that the strongest signal came from the Milky Way, the brightest part of our Galaxy, and especially from the constellation of Sagittarius, which contains the centre of our Galaxy.

This was the first observation of radio waves from space, which became known as 'cosmic radio waves'. It was published in 1932 in the *Proceedings of the Institute of Radio Engineers,* a journal not much read by astronomers, but it was also published as a short note in the more general scientific journal *Nature* with no apparent reaction from the astronomical community. Even when the *New York Times* of 5 May 1933 gave the discovery a full column on the front page, little notice was taken of it. The concept that radio might be useful was outside the vocabulary of even the most astrophysical of astronomers; they were busy interpreting cosmic light, and did not naturally think in terms of radiation with a wavelength more than a million times longer than visible light.

Jansky had done the job of establishing the limits of sensitivity in long-range communications, and the Bell Laboratories had no intention of becoming an astronomical observatory. Jansky turned his attention to other communications work, and did nothing more on cosmic radio noise. It happened that 33 years later the Bell Laboratories were again investigating the basic limitations of long-distance radio communications, but for satellites rather than terrestrial radio links and on much shorter wavelengths, and again there was an unexpected by-product. This was the discovery of the Cosmic Microwave Background (CMB), which will be described in Chapter 9. It is remarkable that these two fundamental discoveries in astronomy were made not in any observatory or university but at a commercial research laboratory.

By the discovery of cosmic radio waves, Karl Jansky became, at the age of only 26, the founder of radio astronomy. He is commemorated in the name of the unit measuring the strength of radio waves from cosmic radio sources. This unit, the Jansky, abbreviated to Jy, appears in practically every one of the tens of thousands of scientific papers in radio astronomy.

Radio waves

Unlike sound waves through the air, and water waves, electromagnetic waves such as light and radio waves need no medium in which to propagate. Any oscillating electric field generates an associated magnetic field, and the magnetic field in turn generates an electric field. The two fields are equal partners, but we usually concentrate on the electric field. Radio waves as we usually encounter them in mobile phones or television are generated by an oscillating electric current in an antenna system mounted on a transmitter tower. They could equally well be generated by an oscillating magnetic source; a good example in radio astronomy is the rotating magnetized neutron star of a pulsar, which is described in Chapter 6.

In a mobile phone the radio signal is generated by an oscillating current in a loop of wire; the same loop is used to receive signals. The pair of short rods, known as a dipole, which may be seen at the focus of a satellite television antenna serves the same purpose, and an elaboration of the dipole is used in the domestic television antennas which sprout from our chimneys. The tiny electric current induced in a dipole by a radio wave is amplified many times by a transistor amplifier. Radio signals from distant satellites, including television programmes, are necessarily weak, but an improved signal can be received using a parabolic reflector with a dipole at its focus. The same principle is used in many large radio telescopes (Plate 1); the parabolic reflector also has the advantage that radio waves are detected almost exclusively from a small range of directions, known as the telescope beam, along the axis of the paraboloid.

Cosmic radio waves are obviously not generated in artificial devices such as loops of wire or dipoles. There are many objects in our Galaxy, and beyond, which contain free electrons in rapid motion, and these can radiate electromagnetic waves at all wavelengths; furthermore, they often conveniently radiate preferentially at radio wavelengths. This is the essence of radio astronomy: celestial objects which are almost invisible to optical telescopes may radiate powerful radio waves, extending the scope of astrophysics and revealing a hidden radio universe.

As already mentioned, radio is at the long wavelength end of a very wide range of electromagnetic radiation which can be received from the cosmos, ranging through infrared, optical, ultraviolet, X-rays, and gamma-rays.[1] The range of wavelengths through this wide spectrum is hard to grasp. Radio wavelengths used in astronomy range from around 10 metres to below 1 millimetre. Visible light is concentrated at much shorter wavelengths, around 500 nanometres (1 nanometre is 10^{-9} metres). X-rays are shorter again, at 1 nanometre or less, but at these short wavelengths the best description of electromagnetic radiation is in terms of photons rather than waves. A photon is a fundamental unit of radiation; any source of electromagnetic radiation transmits or receives radiation in photon units of energy. The energy in each photon is inversely proportional to wavelength; at the short wavelength end of the spectrum, especially for gamma-rays, photons dominate so completely that the radiation is specified only by the photon energy. At radio wavelengths, in complete contrast, we usually only have to consider the wave rather than the photon nature of the radiation (important exceptions are the radio spectral lines discussed in Chapter 3). Furthermore, modern electronic technique allows us to follow the oscillating electric field in a radio wave as picked up by a telescope, and operate on it in ways which are impossible for higher energy radiation, like X-rays and gamma-rays.

In 1932 'short' wavelengths in radio terms meant wavelengths of around 10 metres; Jansky's receiver worked at 15 metres wavelength. Most modern radio communication uses much shorter wavelengths; mobile phones commonly use wavelengths of around 33 centimetres and 16 centimetres. Radio astronomy also has moved to shorter wavelengths, making it far easier to construct narrowly beamed antennas, which we now call radio telescopes.

Treating radio as a wave has a vital influence on the design of radio telescopes. In succeeding chapters we will trace the development of large arrays of radio telescopes, in which the power and speed of electronics and computation have eventually allowed us to combine several thousand telescopes in the SKA, the most powerful telescope of our era.

The First Radio Telescope

Astronomers at first took little notice of the newly discovered cosmic radio waves, and it was left to a gifted amateur, Grote Reber, to make the first maps of the radio sky. Reber was a radio engineer who became an amateur astronomer. In 1958 he wrote:

My interest in radio astronomy began after reading the original articles by Karl Jansky. For some years previous I had been an ardent radio amateur and considerable of a distance communication addict, holding the call sign W9GFZ. After contacting over sixty countries and working all continents, there did not seem to be any more worlds to conquer.

In 1937 Reber built the first reflector radio telescope, 9.5 metres in diameter, entirely with his own hands, and equipped it with receivers working at the shortest possible wavelengths (Figure 2 a, b). After many difficulties, he was able to draw maps of radio emission from the Milky Way using wavelengths of 187 centimetres (160 MHz) and 63 centimetres (480 MHz). Two of his maps are shown in Figure 2c; they are plotted with the North Celestial Pole at the top, so that the Milky Way crosses the map at an angle, as it is normally seen in the sky.

Reber is a heroic figure in the early history of radio astronomy. He worked entirely on his own, using only his own funds. Nothing like his telescope had been built before. It had a sheet metal parabolic reflector surface, on a massive timber framework that could be tilted to any elevation from the horizon to the zenith. The idea was to leave it set at a fixed elevation for 24 hours, and record the strength of the background cosmic noise, originally by writing down meter readings and later by using a strip chart recorder. He did not even explain to his neighbours the purpose of the conspicuous 50-foot high structure that appeared in his back yard in 1937, and he fended off questions from the Chicago *Daily News* by saying:

> There it is, and you can make your own guesses. I'm against talking about inventions before they work out. All I'll say is that it has to do with radio and I don't expect to get rich from it...I hope to have it finished by the end of the month—that is, if I can dodge some of the people who ask me questions about it.[2]

Reber had no astronomical training, so he went to courses at the University of Chicago and eventually interested Otto Struve, the

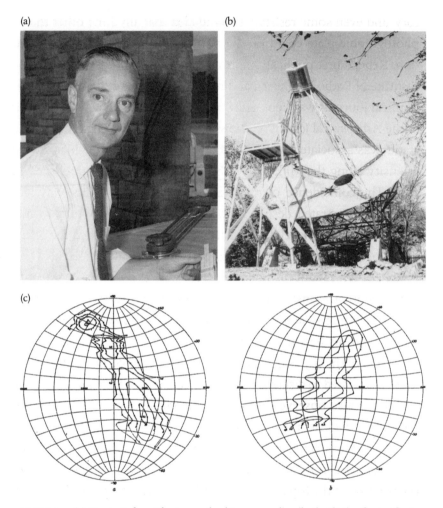

FIGURE 2 (a) Grote Reber. Photographed in 1937, when he built the first reflector radio telescope. (b) Reber's radio telescope. The 31-foot-diameter telescope at Reber's home in Wheaton, IL. *Image courtesy of NRAO/AUI.* (c) The radio sky mapped at 160 MHz. This contour map, published in 1944, shows the intensity of cosmic radio waves concentrated in a strip of sky along the Milky Way. *Image courtesy of NRAO/AUI.*

Director of Yerkes observatory, in his results. Conventional astronomers found it difficult to grasp the significance of his pioneering work; at that time, there was very little electronics in any observatory, and even some resistance to the idea that anything other than a photographic plate could produce any useful astronomical information. Reber's first publication, in 1940, was in the Proceedings of the Institute of Radio Engineers, and his first paper in conventional astronomical journals was kept short on advice from Struve. In this paper in the *Astrophysical Journal* he showed clearly that the origin of 'cosmic static' was associated with the Milky Way. The contour map of the intensity along the Milky Way, shown in Figure 3, was published in 1944. The coincidence of radio and light from the Milky Way was clearly demonstrated, despite the poor resolution of the approximately 12° beamwidth of his telescope, although as we will see later the actual origin of the radio was very different from the visible stars.

Reber then rebuilt his radio receiver using the shorter wavelength of 60 centimetres instead of 1.9 metres, giving a narrower beamwidth. The intensity of galactic radiation falls steeply with decreasing wavelength, and his best maps, published in 1948, show more detail but cover less of the sky. Some features on both maps could later be interpreted as related to discrete sources, which are described in Chapter 5, and Reber also observed radio waves from the Sun, which deserved more than the brief mention he gave it in his papers.

Reber's later career in radio astronomy is a continuing story of adventure and innovation. In 1954 he was observing at low radio frequencies from the top of a mountain in Hawaii, and in 1954 he moved to Tasmania, where he built a huge antenna system for the long wavelength of 150 metres. He is rightly regarded as an inspired pioneer; his radio telescope has been reconstructed and is on display at Green Bank, West Virginia, along with a working replica of Jansky's steerable antenna.

Reber's maps convinced astronomers that cosmic radio waves were worth observing in more detail. A better map covering the whole sky needed larger radio telescopes with more sensitive receivers, and it was not until 1982 that the cosmic radio waves from the whole of the Milky Way were mapped by combining observations made by three radio telescopes, including one in the Southern Hemisphere. Their receivers worked at 73 centimetres wavelength (408 MHz), and the telescope beams were less than 1° across. The result is a spectacular view of the Milky Way (Figure 3 and Plate 2), revealing many features which are invisible to optical telescopes.

This radio map[3] covers the whole sky, arranged so that the Milky Way runs across the centre. As Jansky and Reber found, the most prominent features of the radio sky are the Milky Way, and especially its brightest part at the centre of our Galaxy. This is one of the clearest views we have of our Galaxy: in any optical photograph the Milky Way can only be seen as a patchy belt of starlight. The dark patches that can be seen with the naked eye on any clear night are dust clouds, which obscure much of the light from the Milky Way, and especially the Galactic centre. Dust clouds are no barrier to radio waves.

Besides the Milky Way itself, there are many other novel and previously invisible features in this beautiful radio map, such as the spur reaching up towards the top of the map. But the biggest surprise to the optical astronomers in all these maps is the sheer strength of the galactic radio emission. It is too strong for an origin in stars, although the maps showed that the radio emission and the stars were associated in some way. It was thought at first that the radio waves must come from hot gas filling the space between the stars, but the problem was that no ordinary gas could be hot enough to emit such powerful radio waves. The solution to this puzzle involved very energetic electrons, and an all-pervading magnetic field, which are vital elements of our Galaxy, the Milky Way. Not only our Galaxy, but all others have such a radio source, as was soon discovered for one of our nearest neighbour galaxies, the Andromeda Nebula.

408 MHz

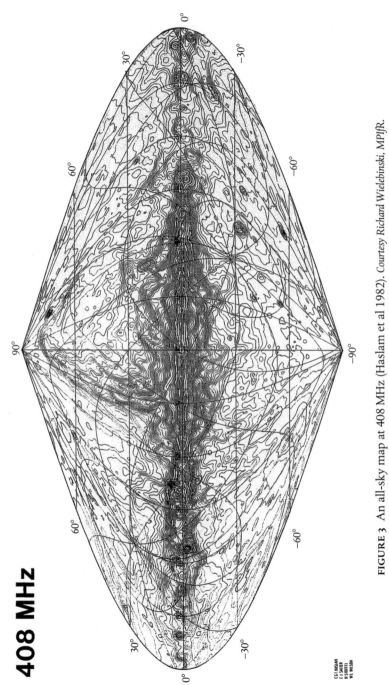

FIGURE 3 An all-sky map at 408 MHz (Haslam et al 1982). *Courtesy Richard Wielebinski, MPIfR.*

The Andromeda Nebula

The Andromeda Nebula (Figure 4) is a typical spiral galaxy, very like our own Milky Way Galaxy. The shape, seen in the optical photograph, is easier to understand than our own Galaxy, because we can see the whole structure from outside. Surely if the Milky Way radiates such strong radio waves, so must the Andromeda Nebula. Indeed it does, and so do other spiral galaxies. The sky is covered with small bright radio sources, like detached patches of the Milky Way. Detecting these distant extragalactic radio sources was an obvious challenge to the early radio observers.

The discovery of radio waves from the Andromeda Nebula came many years after Jansky and Reber first explored the radio sky. Their telescopes were too small to concentrate their beams on a small target,

FIGURE 4 Optical photo Andromeda. *Peter Shah.*

and it was only after World War II that big enough radio telescopes were built. The first, at Jodrell Bank near Manchester, UK, was built for a different purpose, but nevertheless made the vital discovery in 1950. Bernard Lovell, later Sir Bernard, built the telescope. He was a pioneer of radar during World War II, and had returned to Manchester University to use his experience in more constructive and peaceful ends. The idea was to detect cosmic ray showers, using radar to obtain echoes from the cloud of electrons created by the impact of an energetic cosmic ray on the atmosphere.

Lovell first set up a small ground-based radar in a quiet country location 20 miles from the university (Figure 5). He immediately found echoes from the trails of ionized gas left by meteors as they flashed through the atmosphere. The radar worked both by day and by night, even when the sky was clouded over. Seeing meteors in the daytime was exciting, and led to pioneering research on their orbits. But the cosmic ray showers remained elusive.

Lovell decided to build a much more powerful radar, using a large radio telescope antenna looking vertically upwards. This was a parabolic reflector, 218 feet (66.4 metres) in diameter, made by a network of stretched wires (Figure 6). The receiver worked at 2 metres wavelength, and the telescope beamwidth was 2°.

As so often happens when a powerful new scientific instrument is constructed, the most exciting and important results are a surprise. Again no echoes from cosmic ray showers were found, but the telescope turned out to be the most useful instrument yet for observing the cosmic radio waves discovered by Jansky and Reber. Lovell was joined by Robert Hanbury Brown (1916–2002), who already knew of Reber's work and wanted to draw a better map of the radio emission from the Milky Way. The fixed vertical beam of the new telescope was a serious limitation, but Hanbury Brown found that the beam could be swung on either side of the vertical by moving the receiving antenna, which was

FIGURE 5 Meteor radar at Jodrell Bank. This radar was the first instrument to be set up at this famous radio observatory.

mounted on a pole at the centre of the reflector. As the Earth rotated, the telescope beam swept a strip of sky, which could be as much as 15° from the zenith. The radio emission from the Milky Way was easily detected, and the narrow beam of the new telescope also picked out several individual sources within the broad sweep of radio emission in the northern sky. How the Galaxy itself was emitting such powerful radio waves was a mystery, but here were new sources. Were they part of our Galaxy, or possibly might they even be other galaxies like the Milky Way, but at a great distance? The first move was to map the sky round the nearest and brightest galaxy, the Andromeda Nebula.

The historic 1950 map (Figure 7), simple though it is, showed that the Andromeda Nebula generates radio waves with about the same strength as the whole of the Milky Way. Later maps of the Andromeda Nebula, like the one in Figure 8, were made with a much narrower telescope

FIGURE 6 The 218-foot radio telescope at Jodrell Bank. *Central Press Photos Ltd.*

beam; here the spiral arm structure can be seen, and some detailed comparisons can be made between the sources of light and radio.

The Andromeda Nebula and the Milky Way are examples of spiral galaxies. The spiral arms contain young bright stars, but the radio comes not from the stars themselves but from the gas between the stars. Some of the spiral arm emission is from hydrogen gas, but most of the emission from the whole of the Galaxy is closely related to the elusive cosmic rays that inspired Bernard Lovell to set up the new Jodrell Bank Observatory. The connection is explored in Chapter 4.

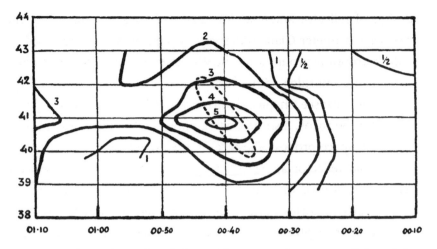

FIGURE 7 The Andromeda Nebula. A radio map made in 1950 with the 218-foot paraboloid radio telescope at Jodrell Bank. *Courtesy of Oxford University Press and the Monthly Notices of the Royal Astronomical Society.*

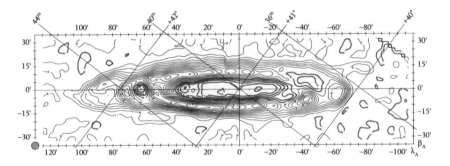

FIGURE 8 A modern radio map of the Andromeda Nebula. A radio map made in 1974 with the 100-metre radio telescope at Effelsberg, Germany. A short wavelength was used, giving the telescope a narrow beam only 5 minutes of arc across. *Berkhuijsen, E. M., Astronomy & Astrophysics, vol 57, page 14, 1977, reproduced with permission © ESO.*

The Big Dishes

Grote Reber's 31-foot-diameter (9.45 metres) dish was the first of a long line of paraboloid reflectors. World War II stimulated the design of reflector antennas for radar, the most notable being the 7.5-metre-diameter Giant Wurzberg, several of which were used by radio astronomy groups after the war (two were used in the interferometer in Cambridge with which I made the first accurate position measurements: see Chapter 5). Space research, which developed rapidly after the 1950s, also needed antennas that could track satellites and deep space probes. The idea of a telescope using a single large paraboloid reflector appealed to optical astronomers, whose large telescopes all used the mirror system introduced by Isaac Newton. Several observatories around the world were introduced to radio astronomy through steerable dishes, which were often mounted like conventional optical telescopes on a polar mount which tracked the movement of stars across the sky. Larger reflectors were mounted differently, steered around the sky on a circular rail track (azimuth movement) and with elevation (altitude) controlled by movement on a horizontal axis. The Lovell Telescope was the first to be constructed in this way, on a so-called alt-az mount. We trace the development of large radio telescopes in Chapter 10, and show in Chapter 11 how very much larger and more sensitive radio telescopes are now being built in the form of arrays of individual telescopes combining to form a huge collecting area.

2

Hot Sun and Cold Planets

Thermal Radiation: the Sun

Everything that is warm radiates; the hotter it is, the more it radiates. An infrared camera images radiation from the warm human body, a hot electric fire radiates red light as well as heat, and the Sun's white light indicates that it is even hotter, at a temperature of about 6000 K. We usually think of the Sun as providing warmth from thermal radiation and light from visible radiation, but the term 'thermal radiation' is used by astronomers to cover the whole wide spectrum of radiation from any warm or hot body. Astronomers are interested in thermal radiation at any wavelength from many kinds of objects, some of which can be studied through their radio emissions even though they cannot be studied optically.

For radio astronomers the Sun is only a minor source of radio waves. Unlike optical astronomy, where bright sunlight makes observing in the daytime practically impossible, radio astronomy can continue throughout 24 hours, since solar radio waves are usually not strong enough to interfere with observations in daytime. Only when there is a solar flare, and even then only at long radio wavelengths, is the Sun overwhelmingly strong. Karl Jansky, the first radio astronomer, and Grote Reber, who followed him in 1935, did not detect the Sun and were able to map the Milky Way through both day and night; the Sun was 'quiet' at the time they were observing. But even when it is quiet the Sun is a very hot body, and like all hot bodies it does emit radio waves. The temperature of a hot body can often be measured by using the strength of its infrared or radio emission as a kind of thermometer, and in a similar way radio can be used to make some interesting measurements of the temperature of the Sun.

The first radio measurements of the temperature of the quiet Sun were made by George Southworth (1890–1972) of the Bell Laboratories. His radio receiver worked at short radio wavelengths, and he found a temperature of 10,000 K, somewhat higher than the 6000 K temperature of the visible Sun (K stands for degrees Kelvin, indicating the 'absolute' temperature scale, with zero at minus 273° Celsius).

Unexpectedly, much higher temperatures were found when other observers worked on longer radio wavelengths; one of the first, Joseph Pawsey (1908–1962) in Australia, found a temperature of 1,000,000 K at a wavelength of 1.5 metres. Why were these radio temperatures so high?

It was soon realized that the radio and light were not coming from the same part of the Sun. The atmosphere of the Sun extends way out beyond the visible surface, forming a *corona* (Figure 9) which can normally only be seen when the Sun is totally eclipsed by the Moon. The corona is at a temperature of over a million degrees; this was the source of the strong long wavelength emission. So the radiation thermometer could be tuned to different parts of the solar atmosphere, long radio wavelengths giving temperatures of 1,000,000 K for the corona, very short radio and visible wavelengths giving 6000 K for the surface (called the *photosphere*), and intermediate radio wavelengths giving around 10,000 K for a transition region known as the *chromosphere*. The corona is effectively transparent to light, so that we see the

FIGURE 9 The solar corona at total eclipse. *Steve Albers, Boulder, CO; Dennis DiCicco, Sky and Telescope; Gary Emerson, E. E. Barnard Observatory.*

photosphere, and opaque to radio, where a radio telescope working at a long wavelength 'sees' the corona and records the high temperature. A radio map of the Sun at long radio wavelengths would not show the normal visible photosphere; instead, it would show a larger diffuse hot disc, completely hiding the normal surface.

All the energy of the Sun comes from a central nuclear furnace, and most of it is radiated away from the surface as heat and light. There is a balance between the energy produced in the core, and the radiation lost from the surface, which determines the surface temperature. How then does the corona get so much hotter than the photosphere? Detailed optical and X-ray photographs of the Sun show a very active surface, with massive turbulence and rapidly changing magnetic fields, all driven by the internal nuclear energy. Some of this mechanical and magnetic energy goes into powerful wave motions, which propagate out and are dissipated in the corona. Plate 3 shows energetic filaments in which there is a strong magnetic field, which is twisted and distorted by the outflow of energy. Changing magnetic fields induce electric currents, which in turn heat the corona. Occasionally there is a sudden catastrophic release of energy, creating a visible bright flare, and radio emission which is very much brighter than even the million-degree coronal radio waves.

Mapping the radio emission from the quiet Sun was a challenging task in early radio astronomy. Only small individual radio telescopes were available, and they had no chance of mapping in sufficiently fine detail. It would be like painting a miniature portrait using only a house decorator's paintbrush. Radio telescopes were in fact only like today's television antennas, with a beamwidth of many degrees. Even to measure the diameter of the emitting solar disc, and compare the corona to the 0.5°-wide photosphere, seemed impossible. The solution to this difficulty, developed in the early days of radio astronomy in Sydney, Australia, and in Cambridge, UK, was to use a radio interferometer.

The Radio Interferometer

The technique of radio interferometry has developed into one of the most powerful tools available to observational astronomy. It is fundamental to the modern multiple telescope arrays, especially the Square Kilometre Array which will be referred to frequently throughout this book. It began with a simple idea: how could a telescope with a broad beam distinguish a small radio source from a large one? Following this, how could one measure its angular size, and its position, with such a blunt instrument?

The Sun has an 11-year cycle of activity; at the peak the visible disc often has several sunspots marking areas of high magnetic fields. Above these spots there are often local regions of high activity, which at times develop into cataclysmic explosions, known as solar flares. The solar flares produce extremely powerful radio waves and streams of charged particles which may be strong enough to affect terrestrial radio communications for a few hours. The active regions are more persistent and may be powerful sources of radio for several days. The source of this sunspot radiation is much smaller than the corona itself. The challenge was to distinguish this concentrated sunspot radiation from the more diffuse radio corona, and to measure its position and its size.

This was first achieved by Martin Ryle (1918–1984)[4] in Cambridge, using a simple radio interferometer consisting of two identical radio antennas connected to a single radio receiver. The way the radio waves picked up by the two antennas add together depends on any difference in the paths they travel; they produce a double signal if the paths are identical, and they cancel if the paths differ by half a wavelength. Figure 10 describes the effect. In Ryle's interferometer the two antennas were spaced by around 100 wavelengths (the wavelength in use at the time was 1.7 metres), so only a small movement of the radio source would change the path difference, and vary the signal at the receiver. The effect on the combined telescope beam pattern is shown in Figure 10c.

As the Sun moved across the sky, the signal detected by the receiver traced a sinusoid on top of a smooth curve (Figure 10e). The concentrated narrow source of the sunspot radiation traced the sinusoid as it crossed the lobes in the combined telescope beam (Figure 10d); this sinusoid is known as the interference pattern of the interferometer. The smooth curve under the sinusoid was the signal from the corona, which was spread over a much larger angular width than the spacing between the lobes in Figure 10.

The concept of this radio interferometer was very close to a classic observation in optical astronomy, in which the angular diameter of a visible star was measured for the first time. This was achieved in 1920 by Albert Michelson (1852–1931), who had already in 1907 received the

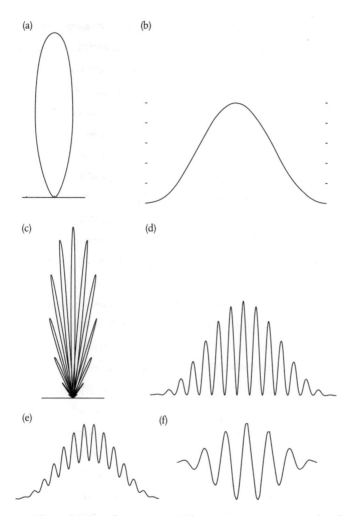

FIGURE 10 The radio interferometer. (a) The reception pattern (polar diagram) of a single antenna. (b) A trace from a radio source crossing this beam. (c) The reception pattern (polar diagram) of a connected pair of antennas, showing an interference pattern. (d) A trace from a compact radio source crossing the interference pattern. (e) The trace from a radio source whose angular diameter is comparable to the spacing between the lobes of the interference pattern. (f) In many interferometers, as described in Chapter 10, only the interference pattern is recorded.

first Nobel Prize to be awarded to an American scientist for the famous Michelson–Morley experiment in special relativity. The angular resolution of an optical telescope depends on its diameter, and the largest available at that time was the Hooker telescope on Mount Wilson, whose 100 inch (254 centimetres) diameter was almost but not quite sufficient to resolve the diameter even of the largest stars. With Francis Pease (1881–1938), Michelson arranged to feed light into the telescope from two mirrors at the ends of a rigid beam mounted on the front ring of the telescope aperture, forming an interferometer with 6-metre spacing. This proved to be just sufficient to measure the diameter of the red giant star Betelgeuse, which they found to be 0.047 arcseconds. With this example in mind, the new radio technique became known as the Michelson interferometer.

This observation showed that sunspot radiation was indeed from a source much smaller than the corona, and that it was localized to the vicinity of the visible sunspot, and furthermore that it was far brighter than any conceivable thermal radiation. We will encounter other examples of this *non-thermal* radiation in later chapters, but the significance of this first radio interferometer is far greater than the discovery itself.

The concept that an interferometer could distinguish between large and small angular diameter radio sources was the first step towards making a complete map of the distribution of radio emission across the Sun, and later across many different kinds of radio source. Eventually interferometers developed into a new kind of radio telescope, through a process known as aperture synthesis. This will emerge through several stages in this book, culminating in the Square Kilometre Array in the final chapter. At this stage we will simply note that the larger the spacing between the elements of an interferometer, the finer the angular detail which the telescope can distinguish.

The Sea-cliff Interferometer

In Australia, Joseph Pawsey and his colleagues also set out to distinguish between small and large sources of waves on the Sun, using a different type of interferometer. Their interferometer, working at a similar long wavelength of 1.5 metres, used a single antenna system directed at the horizon from the top of a high cliff near the entrance to Sydney Harbour. As shown in Figure 11, radio waves from the sun when near the horizon are received both directly and via a reflection in the sea, with the same effect as an interferometer with two antennas separated by twice the height of the cliff. (This technique is analogous to an optical device known as Lloyd's mirror.) The effect is shown in the trace of receiver output in the figure, with a small source tracing a sinusoid while a large source adds a steady background. This same

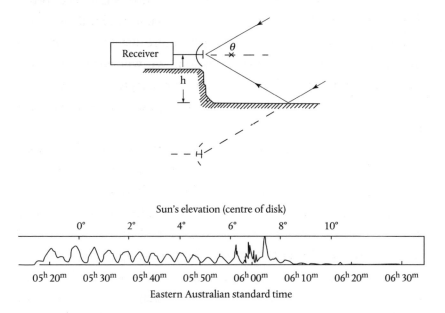

FIGURE 11 Distribution of radio brightness across the Sun: (a) at a long wavelength (2.5 m, 120 MHz), where the corona dominates; (b) at a short wavelength (9 mm, 33 GHz), where the inner part of the corona is seen brightly at the edge of the chromosphere.

interferometer was later used by John Bolton in his detection and location of cosmic radio sources, as described in Chapter 5.

The interferometer concept was later developed in Australia by Wilbur Christiansen, who constructed a line of 32 identical antennas, all connected to the same receiver. The Sun then traversed a set of narrow beams, spaced well apart, which gave a clearer cut across the disc. Using a shorter wavelength of 21 centimetres to obtain a better resolving power, this gave a distribution of radio brightness across the Sun which was markedly different from both the large-diameter corona seen at longer wavelengths, and the disc typically seen optically and at short radio wavelengths, know as the photosphere. The result was a *limb-brightened* Sun, seen in Figure 12.

The interpretation of these observations of the solar atmosphere brought into radio astronomy the concepts of transparency and opacity. The corona is transparent to short wavelength radio, as it is to light, so that radio waves from the photosphere pass through it without any addition from the hot corona. But at long radio wavelengths

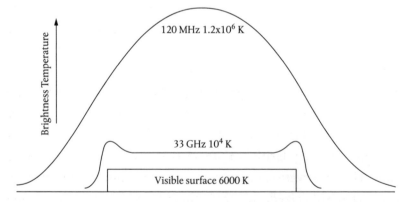

FIGURE 12 Distribution of radio brightness across the Sun: (a) at long wavelength, where the corona dominates; (b) at short wavelength, where only the photosphere is seen; (c) at an intermediate wavelength, where the inner part of the corona is seen brightly at the edge of the disc.

the corona is opaque, and the radio brightness is due to the corona instead of the photosphere. Over a range of wavelengths there is a dramatic change in radio brightness temperature, between 10,000 K at short wavelengths and 1,000,000 K at long wavelengths. In optical astronomy transparency has long been referred to in terms of *optical depth*, and the same term is often used in radio astronomy. A transparent corona has zero optical depth, and contributes nothing to the radio brightness, while the full coronal temperature only shows up when the optical depth is greater than unity. The surface brightness temperature increases linearly with increasing optical depth if the optical depth is small (much less than one), approaching the actual temperature of the gas when the optical depth is greater than one.

A consequence of the large optical depth of the corona at long wavelengths is that other radio sources more distant than the Sun will be obscured if the corona, which is larger than the visible Sun, crosses their line of sight at some time during its annual traverse of the sky. Even the outer parts of the corona can have an effect, since the ionized gas of the corona will bend rays from the distant source out of the line of sight, making the corona an even larger obscuring disc. Fortunately, one of the strongest radio sources in the sky, the Crab Nebula, is in the track of the annual journey of the Sun around the sky. This occultation of the Crab Nebula by the Sun was observed in Cambridge in June 1952, using long radio wavelengths, when the effect of the corona was found to extend to a distance of 10 solar radii. This surprisingly large effect was partly attributable to a scattering effect in irregularities in the electron distribution in the outer corona. This observation eventually had a momentous consequence; Antony Hewish[5] seized on the possibility of observing scattering effects at larger distances from the Sun, and eventually built a very large radio telescope array dedicated to investigating the extension of the solar corona into interplanetary space. It was during the commissioning observations with this telescope in 1967 that Jocelyn Bell saw the first traces of pulsars, a discovery that we commemorate in Chapter 6.

Solar Radio Bursts

Although radio waves from the Sun are usually quiet enough not to interfere with terrestrial radio or radar systems, there is sometimes a spectacular outburst of activity, especially at long wavelengths. The effect was first identified by James Hey,[6] who was one of the pioneers of radar in World War II. Radar operators searching for echoes from aircraft or distant ships have to look for a faint echo against a background of radio noise, some of which may be deliberate jamming by an enemy transmitter. In February 1942 the German battleships Scharnhorst and Gneisenau sailed undetected through the English Channel under cover of jamming transmissions, and Hey was given the task of locating the source of the troublesome transmitters on the ships and on the mainland of Europe. Following up many other operational reports of jamming, he found another unexpected and very powerful transmitter—the Sun.

The Sun occasionally flares into a dazzling radio brightness which can be picked up by any long wavelength radar receiver, even if it is not pointing directly at the Sun. The only activity visible in an ordinary photograph at this time is a sunspot, or a group of sunspots, on the surface. This is the seat of great electrical and magnetic activity, which occasionally breaks out as a huge cloud of energetic electrons and protons shooting out through the corona. This *solar flare* can be photographed in a visible spectral line emitted from hydrogen, as in Plate 4. As these energetic particles travel outward they stimulate radio emission, first in the lowest parts of the corona at high radio frequencies, and then at progressively lower frequencies as the flare activity moves out to higher and less dense parts of the corona. This was the radio transmitter that was jamming the radars which should have seen the German warships. It was an even more effective jammer than any man-made signals, especially because the radio frequencies from the solar flares swept right through the bands used by the early radars; there was no escape by retuning the receivers.

Later observations by Australian radio astronomers, especially Paul Wild,[7] showed how the emitting region of a flare swept in frequency as it travelled upwards through the corona (Figure 13).

The cloud of energetic particles in a solar flare can sweep right through the corona and out into interplanetary space, travelling on past Mercury and Venus to reach the Earth. This is the time, 2 days or so after the original flare, when the northern and southern polar skies are lit by a display of the aurora. Even after their long journey, the protons and electrons of the flare strike molecules high up in our

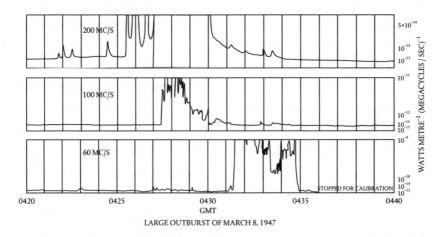

LARGE OUTBURST OF MARCH 8, 1947

FIGURE 13 Solar radio flare. The downward sweep of the radio frequency occurs as the energetic particles of the flare move rapidly outwards in the solar corona. *Reprinted by permission from Macmillan Publishers Ltd: Nature Payne-Scott et al. 1947 'Relative Times of Arrival of Bursts of Solar Noise on Different Radio Frequencies' vol. 160, p. 256. © 1947.*

atmosphere and excite them to glow with their characteristic colours. Wild discovered an even more energetic type of burst, in which the particles travel at one-third the speed of light and reach the Earth after only half an hour. The electrical disturbance may be strong enough to upset the magnetic field of the Earth, and even to induce surges in power lines, sometimes disrupting the power supply over large areas.

Cold Planets and the Moon

The surface temperatures of our Solar System planets are mainly determined by their distances from the Sun. Happily for us, planet Earth is at an intermediate distance where the temperature is suitable for life; Venus and Mercury are too hot, while Mars and the outer planets are too cold. The surface temperatures of planets are known from their radio or infrared emissions; most of these measurements are averages, but for some there have been orbiting spacecraft which have revealed large variations of temperature with latitude and with rotational phase. These are all surface temperatures, averaged over a depth depending on the wavelength; for infrared that means the top millimetre, and for radio as deep as 1 metre, where there is less variation. The coldest is Pluto, the furthest planet (although recently demoted and known as a dwarf planet). Pluto's temperature is 45 K (minus 228 Celsius). On Mercury, closest to the Sun, the temperature varies between 300 K and 440 K.[8]

The Lunar Reconnaissance Orbiter, launched in 2009 and orbiting only 50 kilometres above the surface of the Moon, uses infrared radiometers to measure surface temperatures. Temperatures at the equator range from 400 K in full sunlight down to 100 K at lunar night. At the bottom of one crater near the pole, where no sunlight ever reaches, the temperature is only 25 K.

The temperatures on some planets are not so easily specified. Venus, in particular, is notorious for possessing an atmosphere of carbon dioxide and sulphuric acid, which acts as a blanket, heating the planet by a greenhouse effect and obscuring the surface for most wavelengths. Only at very short radio wavelengths, down to 1 millimetre, is the surface seen to have a temperature of about 300 K, not much hotter than Earth. At the longer wavelength of 10 centimetres only the blanketing atmosphere is seen, with a temperature of 650 K.

The outer planets, Jupiter, Uranus, and Neptune, do not have a solid surface. As in the case of Venus, temperatures measured at different wavelengths can be quite different: Uranus for example is hotter at 10 centimetres than at 1 millimetre, but at shorter infrared wavelengths, which penetrate deeper, the temperature is much greater; Uranus is suspected to have an internal source of heat. Saturn also has a complicated atmosphere with temperature apparently varying from around 140 K at short radio wavelengths to 400 K at 1-metre wavelength. Radio emissions from these outer planets have been mapped in some detail, showing enhanced emission from an equatorial belt (Figure 14). This extra emission has a different character from the thermal emission from a hot surface; it is attributed to synchrotron radiation from high-energy electrons, a mechanism which we explore in a different context in Chapter 4.

Jupiter, as observed at low radio frequencies, exhibits some spectacular emissions. On a visiting Fellowship to the Carnegie Institute of Washington in 1953, I collaborated with Bernard Burke[9] in constructing a large telescope array working at 22 MHz (14-metre wavelength). When I left to return to Cambridge, this radio telescope was left set at an elevation where it would observe the Crab Nebula every day. Quite by chance, Jupiter also crossed this line of sight every day,

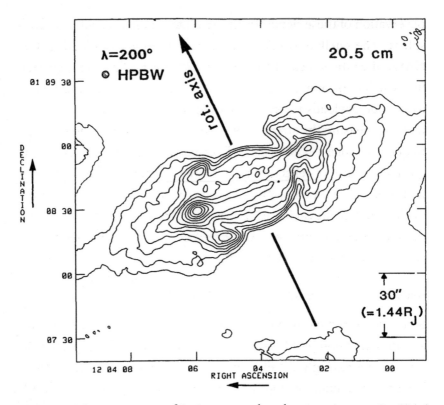

FIGURE 14 A contour map of Jupiter at wavelength 20 centimetres. *Republished with permission of Annual Review of Astronomy and Astrophysics, from 'Radio images of the planets', de Pater, I., Vol 28, 1990; permission conveyed through Copyright Clearance Center, Inc.*

and Burke and his colleague Kenneth Franklin observed a large and rapidly fluctuating signal from the planet.[10]

This unexpected radiation originated high in the planet's atmosphere, from high-energy electrons gyrating in Jupiter's strong magnetic field.

Planet Earth itself, as seen from outside, would be observed to radiate powerful radio waves, but at lower frequencies again, due to resonances in the terrestrial ionosphere. I was involved in the first observation of these signals from a satellite orbiting outside the iono-sphere. The satellite was Ariel II, the second of a series launched by NASA after the shock of the first Soviet satellite Sputnik I, and the first satellite to carry a radio astronomy receiver. The UK was invited to provide payloads for six of this series, and I proposed an experiment to measure the strength of cosmic radio waves in the very low frequency range 1–3 MHz, which could not be observed from the ground because of reflection from the ionosphere. The antenna system was a wire dipole 40 metres long, deployed from the spinning satellite by centrifugal force. With my student Jan Hugill, we achieved the measurement, but we also observed the intense resonant radio emissions at around 1 MHz.[11]

Radar: Meteors, the Moon, and the Planets

Some of our best pictures of the surfaces of Mars, Venus, and Mercury are not true photographs, but maps made by radar from spacecraft. These are amazingly detailed, with very accurate measurements of heights. Spacecraft operating from only a few hundred kilometres above the surface of a planet have an obvious advantage over radars on Earth, perhaps more than a thousand times further away. Nevertheless, some amazingly detailed radar observations of the Moon and the planets were made from ground-based radars before the space age, and there is an interesting history of early radar observations from before the era of space research.

The first astronomical objects to be studied by radar were meteors. The echoes from these were often seen in the defence radars set up round the coast of Britain; they were surprisingly strong, considering that meteors are no more than tiny pieces of dust, but the echoes were actually from the long trails of ionized gas which the meteors created as they crashed into the atmosphere. The trails were typically some tens of kilometres away, so that the radar pulse took about a thousandth of a second (a millisecond) to travel to the meteor trail and back.

The next target was the Moon. This was out of range for most war-time radars;[12] the travel time to and from the Moon is 2.6 seconds, and the large distance meant that sensitivity had to be increased at least 1000 times. One way of increasing sensitivity is to make many observations and add the results, so cancelling out some of the noisy background which hides the echo. This technique was used in 1946 by Zoltan Bay, an enterprising Hungarian scientist who has the credit for the first radar echo from the Moon.[13]

The US Army Signal Corps followed, using a more powerful trans-mitter, and in Australia echoes were obtained by using the 100-kilowatt transmitter of a radio broadcasting station. With modern radars, and large steerable radio telescopes, the Moon is now regarded as an easy target. The next challenge was to get a radar echo from the planet Venus. This was achieved in 1961, when Venus was at its closest approach to the Earth.

It took 15 years of development to achieve the step between the Moon and Venus. An improvement of sensitivity by a factor of over a million was needed. Remarkably, this was achieved almost simultane-ously in several countries, but eventually planetary radar became the province of large and powerful radars in the USA, such as the Haystack radar. The most spectacular of these is at Arecibo in Puerto Rico, using a reflector telescope 1000 feet (300 metres) in diameter—the largest parabolic reflector in the world (see Chapter 10). Measuring the time for a radar pulse to travel to Venus and back gave a dramatically improved value for the scale of the whole solar system.[14]

A powerful Earth-based radar giving strong echoes from the Moon and the planets can make surprisingly good maps of the surface, even though the beam of the radar is itself not fine enough to scan the sur-face. Two tricks can be used, called Delay and Doppler. Delay divides the echo in time, the first part coming from the nearest part of the surface of the spherical target, and the later parts from nearer the edge. Doppler depends on the rotation of the planet: one side comes towards the radar, and the other away, so that the echoes from the two sides can be distinguished. Even the Moon shows this Doppler effect, since it oscillates to and fro by as much as 10°. The combined techniques of Delay and Doppler have produced very good maps of the Moon and the planet Venus.

The most detailed maps of planetary surfaces are made by photography from orbiting spacecraft, but the radar observations are still valuable since they probe the surface to a depth of several wavelengths. Radar is also particularly useful because of its ability to measure velocities via the Doppler effect; it was a radar measurement that established in 1965 that the spin and orbit of Mercury were locked in a 3:2 ratio, three rotations occurring in every two orbits round the Sun.[15]

Distances Measured from Spacecraft

The most important quality of radar, compared with passive reception of radio, is the accurate measurement of distance. The time taken for a pulse to reach a planet and return, given that the pulse travels at the speed of light, measures the distance to the target. Corrections may have to be made for delays through the atmospheres of the Earth and the planet, but a far more important correction is due to a relativistic effect, the Shapiro delay. This is an increase in travel time when a ray path is in the gravitational field of a massive object, in this case the Sun. For Venus the effect is greatest when the line of sight to the planet is closest to the Sun, when it amounts to 200 microseconds. Irwin Shapiro (b. 1929) predicted this in 1964; the first measurements, in 1966/7 at the MIT Haystack radar, confirmed the prediction within 10%.[16]

Many such measurements have now been made to great accuracy, including propagation delays in transmissions from transponders in spacecraft observed with a line of sight close to the Sun. The effect has also been observed in timing observations of binary pulsars (see Chapter 7).

The scale of the solar system is specified in terms of the Astronomical Unit, which is the mean distance of the Earth from the Sun (the Earth's elliptical orbit requires a careful interpretation of this definition). Combining the best radar and spacecraft measurements now gives a value for this unit, the a.u., of 149,597,870,700 ± 3 metres. The accuracy is remarkable: it takes light or radio only one hundredth of a microsecond to travel the quoted distance of 3 metres.

3

Our Galaxy: the Milky Way

The Hydrogen Line

Astronomy in the Netherlands was almost at a standstill in World War II, during the German occupation. At considerable personal risk, and under very difficult circumstances, Professor Jan Oort (1900–1992) in Leiden managed to keep a small research group at work. Oort also obtained copies of the *Astrophysical Journal*, published in America, and he was one of the few astronomers who took serious note of the radio work of Jansky and Reber. This was discussed at one of his seminars, held secretly in a basement, out of which came a momentous prediction. Oort was interested in the structure of the Milky Way galaxy, and particularly the dynamics of its rotation. The featureless spectrum of Jansky's radio was of little use in elucidating the structure of the Galaxy; it would be far more useful if there were a spectral line to be observed, whose structure would reflect the dynamics of its source. Throughout the radio and optical spectrum, every kind of atom and molecule radiates at well-defined wavelengths; for example, the familiar yellow colour of sodium street lights is due to a spectral line in which electrons in individual sodium atoms jump between fixed energy levels. Figure 15, which shows how spectra in visible light and in radio are presented, shows why these sharply defined features are called 'lines'.

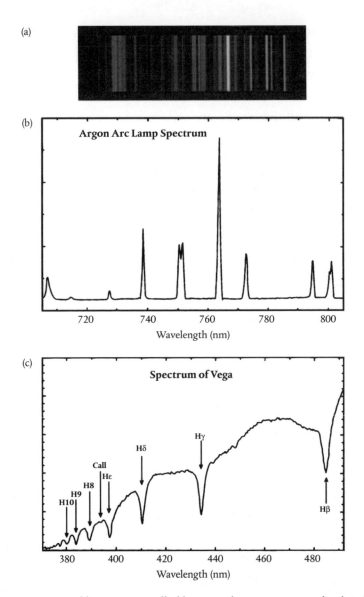

FIGURE 15 Spectral lines are so-called because they appear as such when light is split by a spectroscope, as in the spectrum of light from an argon arc (a). A spectrometer shows the lines graphically, as in (b), but narrow features are still referred to as 'lines'. Spectral lines may be seen in absorption, as in (c) which shows part of the spectrum of light from the bright star Vega. The so-called Balmer series of lines here are due to absorption in hydrogen. *Ian Morison, Jodrell Bank Observatory.*

Oort knew that hydrogen atoms were common as a dilute gas throughout the Galaxy, although they did not show themselves by emitting any visible spectral lines. He asked a student, Hendrik van der Hulst (1918–2000), to look at the structure of hydrogen atoms, and see if they could radiate a radio spectral line. The question was:

> Is there a spectral line at radio frequencies we should in principle be able to detect? If so, because at radio wavelengths absorption should be negligible, we should be able to derive the structure of the Galaxy. We might even be able to detect spiral arms, if they exist.

Van der Hulst soon found that neutral hydrogen, which is abundant in interstellar space, has a prominent spectral line at a wavelength of 21 centimetres, and this was duly observed in 1951, first in the USA and soon afterwards in the Netherlands and in Australia. The three observations were published together in the same issue of *Nature*, which was an act of courtesy by the discoverers, Harold Ewen (b. 1922) and Edward Purcell (1912–1997). Oort and his receiver expert, Lex Muller, would have been the first to detect the line, but they were delayed at a crucial moment when their equipment caught fire.

Spectral lines are formed in a quantum process, in which the energy of an atom or molecule changes between two discrete levels. The transition which van der Hulst found in neutral hydrogen is between two states with only a small difference in energy, so that the spectral line occurs at a low radio frequency rather than in the visible, infrared, or ultraviolet spectrum.

The hydrogen atom has a nucleus containing a single proton with one electron in orbit round it. Atoms usually emit visible spectral lines when an orbiting electron jumps from one fixed orbit to another, but in the cold depths of space within the Galaxy the electrons stay only in the lowest orbit. There are however two states of the hydrogen atom with the electron in the lowest orbit. Both the proton and the electron have another quantized property called spin, which we visualize conveniently as like the spin of tops. The spins of the proton and electron in a single isolated atom are either aligned or opposed. The energies of these two states differ by a minute but precisely defined amount. A change from the unaligned to the aligned configuration releases a quantum of energy as a photon, with a very precisely determined frequency.

This spontaneous transition is very unlikely; in spectroscopy and quantum mechanics it is even called a forbidden transition. Hydrogen atoms in interstellar space lead a lonely and isolated existence, and a completely isolated neutral hydrogen atom in the more energetic state would stay as it is for millions of years before making this transition. But van der Hulst pointed out that there is so much hydrogen in the Galaxy that the spectral line might nevertheless be observable, and so indeed it was. The spectral line occurs at a wavelength of 21.106 centimetres (a frequency of 1420 MHz), and it only required a small extension of receiver technique to test the idea.

The Spiral Structure of the Galaxy

Apart from the excitement of being able at last to detect hydrogen in the Galaxy, and the triumph of the successful prediction in such circumstances, the value of the 21-centimetre line observations lay, as Oort had suggested, in a miraculous key to the dynamics of the Galaxy. Up to that time the structure had been investigated through measurements of the velocities of stars rather than the gas between them. Velocities are found from precise measurements of the wavelengths of the abundant spectral lines in starlight. Any velocity in the line of sight shifts the centre frequency of a spectral line towards shorter wavelengths for an approaching source and towards longer wavelengths for a receding light source. This is the Doppler effect, familiar in sound waves from the changing note of a siren in an emergency vehicle as it drives past at high speed. The amount of wavelength shift depends on the ratio of the speed of the vehicle to the velocity of sound; for the stars it is the ratio to the velocity of light. What Oort and others had been after for many years was to find a pattern in the Doppler shifts from stars distributed throughout the Galaxy, looking for a pattern which would reveal the organized rotational motion of stars on the large scale. This was difficult; distances to stars are not well known, and individual stars have their own velocities, which confuse a search for an organized pattern. Hydrogen gas between the stars gives a much clearer picture; its only bulk motion is due to the rotation of the Galaxy. The 21-centimetre line is narrow, and Doppler shifts are easily measured.

By 1958 a map (Figure 16) was available of the distribution of hydrogen in the plane of the Galaxy.[17] This famous map, which combined observations made in the Netherlands and in Australia, extended over the whole of the galactic system, and gave the first picture of spiral structure in our Galaxy. The observations along any line of sight give only the distribution of velocities, and velocities have to be turned into distances by using a model of the rotation of the Galaxy. Oort and his collaborators assumed that the stars and gas were in circular orbits, moving at speeds which were related only to the distance from the centre of the Galaxy. The speed at any radial distance would depend only on the gravitational pull of the matter inside the orbit. The map is centred on the galactic centre, and the radius from the centre to the Sun is upwards in the map. In the directions of the centre and the anti-centre, on a vertical line in the map, the circular motion is across the line of sight, so there is no Doppler shift from circular motion, and no distances can be found.

The variation of speed with distance[18] from the centre, the 'rotation curve', which emerges from the analysis along with the hydrogen distribution, is particularly valuable because it reveals the distribution of

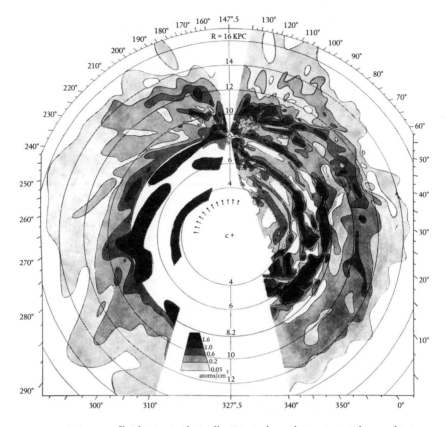

FIGURE 16 A map of hydrogen in the Milky Way galaxy, drawn in 1958 by combining hydrogen spectral line observations made in the Netherlands and Australia. This was the first map to show the spiral arm structure of our own Galaxy. *Courtesy of Oxford University Press and the Monthly Notices of the Royal Astronomical Society.*

mass with radial distance from the centre. Figure 17 shows the curve, incorporating some detail near the centre and extending to radii well beyond the solar distance at 26,000 light-years. At large distances from the centre, nearly all the visible mass in the Galaxy is inside the orbit, so that the velocities should behave like velocities of planets in the solar system, falling with distance according to Kepler's laws. Instead the velocity curve becomes flat, and stays flat as far as it can

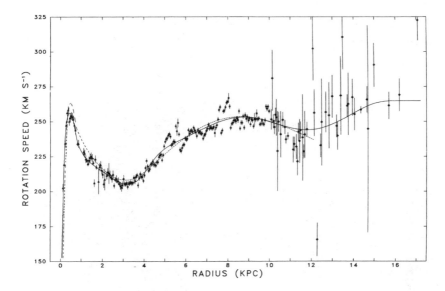

FIGURE 17 The rotation curve of the Milky Way. The curve, which is found from measurements of velocities of stars and gas, shows how the rotation increases from the centre out to a distance of around 15,000 light-years, and continues at more than 200 kilometres per second as far as measurements can be made. *Clemens 1985. Reproduced by permission of the AAS.*

be measured. This phenomenon is seen also in the rotation curves of other galaxies, which extend without diminution out beyond the visible discs delineated by stars. The inevitable conclusion is that the gravitational force is greater than that due to the visible matter, and must be due to some invisible 'dark matter'. We will encounter this mysterious extra constituent of the Universe again in a later chapter on cosmology.

The spiral structures of other galaxies are familiar from the photographs which are taken by every newly constructed optical telescope to prove and advertise its performance. The rotational dynamics of many have been mapped by spectroscopy both by the light of stars, and by radio for their interstellar hydrogen. Their spiral structures and their rotation curves show the same general features, including the high velocities in the outer parts which indicate the presence of dark matter. There are convincing theories of the formation of spiral arms as density waves travelling outward, controlled by gravity, which reproduce the observed patterns of loosely and tightly wound arms. But there are many details which do not conform to the simple pattern. The spiral galaxy M51, known as the Whirlpool, which is often quoted as an example of spiral structure, is far from simple (Plate 5). Close to it, at the end of an extended spiral arm, is a separate galaxy which has a significant gravitational pull. This situation is now recognized as an example of a common phenomenon: a gravitational interaction, almost a collision, between pairs of galaxies.

The Centre of the Milky Way

The dust clouds which lie in the plane of our Galaxy completely cut off our view in visible light of the galactic centre. The longer wavelengths of infrared light, and the whole spectrum of radio waves, penetrate the dust clouds and give us a view of a cauldron of activity which has been hidden from us. The radio maps of other galaxies give us some idea of what to expect, although at such great distances it is impossible to see sufficient fine detail.

Figure 18 is a map of a 50-centimetre-wavelength radio emission from the Andromeda Nebula. This is a map made in 1984 using the Westerborg Synthesis Radio Telescope in the Netherlands,[19] an early example of the high angular resolution obtained by aperture synthesis (see Chapter 10). Here we see two main features which would be difficult to discern from a position like ours in our Galaxy, immersed as we are in the disc. This nebula is tilted out of our line of sight, so the ellipse in the map is in fact a roughly circular ring at a distance of about 30 light-years from the centre, similar to the distance of the Sun from the centre of our Galaxy (Plate 6). Inside the ring there is little emission, until close to the centre (and unresolved in this map) there

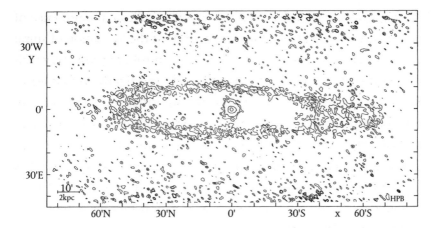

FIGURE 18 Radio map of the Andromeda Nebula, showing a bright source at the centre. An early example of a map made using an aperture synthesis telescope, the WSRT. *Bystedt, J.E.V. et al., Astronomy & Astrophysics, vol 56, page 277, 1984, reproduced with permission © ESO.*

is a region of bright emission. There seems to be a similar concentration of emission near the centre of our Galaxy; although it is hard to delineate, it is evidently several degrees long and tilted across the line of sight, forming an inner connection with the spiral arms. This is the so-called bar, a structure common to many spiral galaxies. The most interesting features at the very centre are far smaller, and can only be delineated by using telescopes with an angular resolution of 1 arcminute or better. Their complexity can only be unravelled with a combination of radio, X-ray, and infrared telescopes.

Radio astronomy has now penetrated the centre of our Galaxy in far greater detail than is possible for extragalactic nebulae. The suspicion that there was something extraordinary at the very centre was, however, inspired by observations of some radio galaxies whose exceptionally very energetic emissions came from a concentrated central region known as an Active Galactic Nucleus, or AGN. In 1964 Zel'dovich[20] proposed that this activity was associated with a massive black hole. The energetic process was supposed to be the release of gravitational energy as stars and gas fell into the black hole. The Milky Way certainly did not have an AGN at its centre, but it was already known that there was a compact radio source in the constellation of Sagittarius, close to the centre of rotation. In 1969 Donald Lynden-Bell suggested that this might be a black hole, albeit much less active than the AGN of a radio galaxy.

The 1970s was a period of intense activity at the radio observatories. The possibilities of constructing radio telescopes with very high angular resolution was being explored and exploited, using new techniques of interferometry and aperture synthesis. A single reflector, or 'dish', is a very blunt instrument for mapping a complex source, as indeed the Galactic centre proved to be. Most of the activity is within a volume only a few light-years across, at a distance of some 25,000 light-years, so a map covering an angle of only about half an arc-minute was needed, with detail down to an angular resolution of 1 arcsecond or better. The beamwidth of a single dish is measured as the ratio of wavelength to diameter; even at the short wavelength of 1 centimetre the largest dishes, about 100 metres across, have a beamwidth of half a minute. But, as we will see in Chapter 10, the ideas of interferometry were now being explored, and the rapidly expanding capabilities of digital computers were being applied to the use of long base-line interferometers for map making by aperture synthesis. In 1974 NRAO, the National Radio Astronomy Observatory in the USA, started using an interferometer with a base-line of 34 kilometres at 3.5 centimetres wavelength, and Bruce Balick and Robert Brown[21] were able to show that the Sagittarius complex contained a source within an area of 1 by 3 arcseconds. A good optical telescope would do well to beat that sort of angular resolution, but this observation went further: it showed that the source contained three separate objects, one of which was less than a tenth of an arcsecond across. At the distance of the galactic centre, that object must be less than one light-day across, not much larger than our Solar System.

This central object became known as Sagittarius A, or Sgr A, and the most concentrated object within it was designated Sgr A*. Full scale mapping of the whole region followed, using the new synthesis telescope arrays at Westerbork (the WSRT) in the Netherlands and the Very Large Array (the VLA) in the USA. As can be seen from Figure 19 and Plate 6, the whole area is filled with radio sources of several kinds. Four or more of these, marked SNR, are remnants of supernovae. Along the centre line there are small clouds of emission, which are the hot hydrogen regions surrounding energetic young stars. Sgr A is the overexposed object in the centre. Surrounding it are some remarkable filaments, labelled as the Radio Arc and the Non-thermal Radio Filaments (the NRFs). Further west are more filaments, including the Snake. The Radio Arc and the filaments contain high-energy electrons emitting radio as they move in a magnetic field (synchrotron radiation, see Chapter 4);

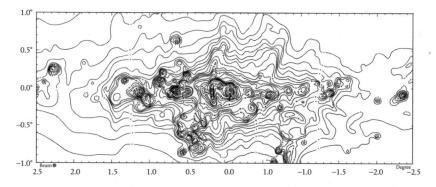

FIGURE 19 Radio map of the centre of our Galaxy, using a high resolution aperture synthesis radio telescope. *Altenhoff, W.J. et al., Astronomy & Astrophysics, vol 35, page 30, 1979, reproduced with permission © ESO. Astronomical Journal 119:207.*

they are interpreted as jets expelled from near Sgr A*. The energy in this pattern of activity is far short of the Active Galactic Nuclei seen in many distant galaxies, but it did suggest that Zel'dovich's prediction of black holes as central energy sources might apply to our own Galaxy.

Sgr A* has now been observed over the whole radio spectrum from 1 to 230 GHz (wavelengths 30 centimetres to 13 millimetres). The X-ray satellites Chandra and Newton both observed it as an unresolved source, but found that it was very variable on a short time scale: in one episode in 2001 it brightened by a factor of 45 and faded within only 3 hours; this could only happen if the source was physically very small, less than the smallest planetary orbit in our Solar system.

A black hole by its very nature cannot emit electromagnetic radiation of any kind, so these emissions could only be interpreted as indications of some concentrated source of energy, such as gravitational energy released outside a black hole as gas clouds fall into it. It was, however, clear that there was an extraordinary concentration of mass within a few light-years. The full story was revealed by observations in another part of the electromagnetic spectrum, the infrared.

Although scarcely any visible light reaches us from the galactic centre due to obscuration by dust clouds, infrared light with wavelengths of 1 micron and longer, i.e. more than twice the wavelength of visible light, can penetrate the clouds, allowing a conventional telescope to image individual stars. Two large telescopes have been looking at stars in Sgr A since 1995; the 10-metre diameter Keck in Hawaii and one of the array of four 8-metre telescopes known as the Very Large Telescope (the VLT) of the European Southern Observatory in Chile. A very high angular resolution is needed. The beamwidth of an 8-metre telescope at a wavelength of 2 microns is ideally less than 0.1 arcsecond, which turned out to be small enough to see a remarkable collection of stars close to SgrA*. But this was not easily achieved, since atmospheric turbulence usually smears out any image to a size of at least 1 arcsecond. Refraction in the air above the telescope distorts the wavefront entering the telescope, and the image of a star is constantly moving and becoming blurred, changing on a time scale of less than a tenth of a second. The amazing new technology of 'adaptive optics' allows the distortions to be measured and corrected continuously on this very short time scale. It achieves this by creating an artificial star, measuring the shape of its image, and distorting a small secondary mirror to restore a perfect wavefront. The artificial star is created by a laser beam which excites sodium atoms in a small patch of the upper atmosphere. Star images as small as 0.04 arcseconds have been achieved.

No light, either visible or infrared, has been detected from SgrA* itself. The infrared photographs show only a large concentration of stars within a few light-years of the invisible black hole. It was soon evident that some of these were very fast-moving, at more than 1000 kilometres per second, and that they are in orbit around the black hole. One of them, designated S2, has completed an elliptical orbit in only 15.8 years. Continued observations have produced Figure 20, which shows the elliptical orbits that have been determined for 20 stars. This is an amazing achievement; the whole of these plots covers an area only 1 arcsecond across. All the orbits would fit into a space no larger than the distance from the Sun to the nearest star. Simple Newtonian dynamics can now be applied, telling us that the gravitational field in which these stars are orbiting is that of an object with a mass of about 4 million times that of our Sun, the whole mass being concentrated in a volume inside the closest approach of the smallest orbit. A concentration of 4 million individual stars would soon collapse into a black hole, so we are safe in assuming that the central object in our Galaxy is indeed the black hole which has been sought for since Lynden-Bell's prediction.[22]

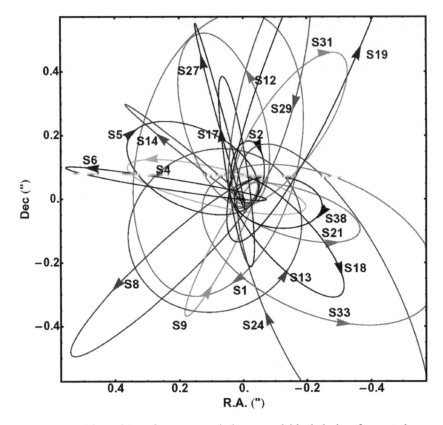

FIGURE 20 The orbits of stars round the central black hole of our Galaxy.
Reproduced by permission of the AAS.

Radio provided the first images of SgrA*, and X-rays discovered its
dramatic variability. The stellar orbits were seen and measured in the
infrared. Has radio any further role to play? Indeed it has. The jets and
clouds surrounding SgrA*, whose synchrotron radiation can best
be mapped by radio, are an indication of activity which can only be
guessed at. Radio observations must, however, be made at short wavelengths, because of the scattering which occurs in the ionized clouds
on the plane of the Milky Way, between us and Sgr A*. The best
chance is with the millimetre wavelength array ALMA, which will

have sufficient angular resolution at wavelengths below 1 millimetre. Observing the activity round our own black hole will provide a model for the multitude of much more distant, and more active, radio galaxies which we explore in Chapter 5.

Hydrogen and Electrons

'I got plenty of nuthin.' So sang Porgy in *Porgy and Bess*. He could have been singing about our Galaxy, the Milky Way. The stars are light-years apart, apparently with nothing in between. But not quite nothing; there is a thousand times more matter per cubic metre in that interstellar space than there is further out in the Universe, in the space between the galaxies. On average, in our part of the Galaxy every cubic metre contains matter equivalent to some hundreds of hydrogen atoms. Most of it actually is hydrogen; and most of that is ionized, split into a proton and an electron. Close to the galactic plane, and particularly in the molecular clouds, a small proportion is in the form of heavier elements, some of which are combined into molecules such as ammonia, NH_3, and carbon monoxide, CO. We have see in Chapter 4 how these molecules reveal themselves by radiating at wavelengths in the radio spectrum; we now look at the way radio explores the ionized interstellar gas which extends throughout the Galaxy. This is possible because the free electrons influence the propagation of radio waves far more than light.

In a vacuum all electromagnetic waves—radio, light, gamma—travel at the same speed of 300,000 kilometres per second.[23] Adding electrons makes a difference to the velocity because they respond to the electric field of the wave. Their response depends on the frequency of the oscillation in the wave. The response is negligible for high frequency waves such as light, but the electrons do respond strongly to radio waves, in which the oscillations are very much slower. The effect on the velocity of a radio wave is small, but it can be measured in the time it takes for a radio pulse from a pulsar to reach us after a long journey through interstellar space.

Pulsars, which are the subject of Chapter 6, radiate short radio pulses, some as short as a millisecond or less, simultaneously over a wide band of radio frequencies. These pulses are ideal for measuring the total electron content of the line of sight between us and the pulsar. When we receive a pulse after it has been travelling for thousands of years, the pulse has been delayed by a small amount. The delay may be only a second or less, which seems a ridiculously small amount to measure, especially when we have no means of timing the start of its long journey. The delay is, however, different at different radio frequencies, and it can be measured by comparing the arrival times at two or more radio frequencies. The effect can be seen in Figure 21. Here the arrival time of the pulse is displayed for 64 adjacent radio-frequency bands. Each line of this figure covers the interval between consecutive pulses, which for this pulsar is 455 milliseconds. The pulse arrives later in successively lower frequency bands, an effect known as *dispersion*. The whole delay at the centre frequency of this plot is 1.2 seconds, an amount which depends directly on the electron content of the line of sight.

This dispersion delay occurs for every known pulsar, in all directions in our Galaxy, showing that there is ionized hydrogen throughout the whole of the space between the stars. Astronomers now use the term *interstellar medium* rather than *interstellar space*. The pulse delay gives us the *total* electron content in a line of sight; to deduce the density we need to know the distance of the pulsar. Fortunately we do know the distance for a considerable number of them, so we can make a three-dimensional map of the electrons in the Galaxy. We can then reverse the process for pulsars whose distance is not known, by measuring a dispersion delay and using the map to find a useful distance.

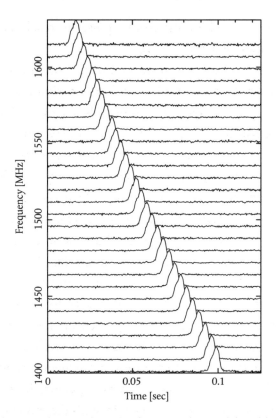

FIGURE 21 The arrival time of a pulse from a pulsar is delayed by electrons in interstellar space. The delay is greater for lower frequencies, as seen from the arrival time in 28 adjacent frequency bands. From 1620 to 1400 MHz the delay increases by 80 milliseconds. *Patrick Weltevrede, Jodrell Bank Observatory.*

Where Are the Electrons?

The general distribution of electrons in our Galaxy has three components, designated by their locations: Galactic Plane (including the spiral arms), thick disc, and Galactic Halo. Left to themselves, the electrons of the interstellar medium would combine with protons to form neutral hydrogen; this is a slow process, as collisions between protons and electrons are rare, but time scales are long enough for this to occur. The interstellar medium has to be kept ionized by some sufficiently energetic process. It was thought for a time that this would be due to the energetic particles of cosmic rays, but it now seems likely that most of the ionization is due to ultraviolet light from energetic stars. This is certainly the case for interstellar hydrogen close to the galactic plane; some stars are powerful enough sources of ultraviolet photons to create spherical regions round themselves in which all the hydrogen is ionized; these stars are young bright stars located in the spiral arms. As seen from the Sun, the spiral arms, with their complex mixture of neutral and ionized hydrogen, and molecules, form a flat disc seen edge-on, so this component of the electron distribution makes a narrow band across the sky within the Milky Way.

The most important electron component of the Galaxy is on a much larger scale. It forms a thick disc extending through the whole plane of the Galaxy between the Sun and the centre, with a thickness about one-fifth of its diameter. On this larger scale the ionization is still due to diffuse ultraviolet light from distant energetic stars located in the plane of the Galaxy. The distribution of electrons through this disc is far from smooth, especially near the plane; an obvious complication is due to supernova explosions, which send waves of ionized gas flowing out from locations in the spiral arms into the thick disc. On average, the electron density in the disc, at its peak close to the plane, is about 10,000 electrons per cubic metre, falling above and below the plane on a scale of 1 or 2 kiloparsecs (the distance from the Sun to the centre of the Galaxy is 8 kiloparsecs, or 26,000 light-years). The thick disc merges into a hot, low-density halo which extends round the whole Galaxy.

Free-free Radiation

Although the pulse delays are the best probes of these thin ionized clouds, there is another way in which the electrons make their presence felt: they can radiate radio waves. This occurs when an electron, as it wanders through the interstellar medium, encounters a proton, to which it is attracted by the opposite electric charge. Whenever an electron is accelerated, it radiates. Unlike spectral line radiation, the electron is at no stage bound to the proton, so this radiation is called free-free. It occurs throughout the radio spectrum, but at low frequencies (metre wavelengths) it is much weaker than the synchrotron radiation which dominates the sky. It is important at millimetre wavelengths, where it has to be distinguished from the cosmic microwave background (see Chapter 9). Free-free radiation depends twice on density; electrons must collide with protons (of which there is the same number), so that the radiation is proportional to the square of the electron density. It therefore shows up any concentration of electrons, and it is much brighter than expected from a smooth thick disc. Close to the galactic plane the thick disc must be very irregular and clumpy, becoming smoother as we look further away from the energizing hot stars and supernovae.

There is no obvious boundary to the thick disc, and we can only conjecture that there is a very thin ionized gas pervading the space between galaxies. This will undoubtedly be explored when we are able to detect pulsars in extragalactic nebulae. So far only the Magellanic Clouds, which are part of the local group of galaxies, are accessible in this way.

Our magnetic Galaxy

When in 1968 we heard at Jodrell Bank Observatory about the discovery of pulsars, Andrew Lyne and I realized that the extremely intense radio pulses must be polarized; furthermore, that the Lovell Telescope at Jodrell Bank was ideally equipped to test the idea. In a radio wave, the oscillating electric field is always across the direction of travel, and if it is polarized the electric field, as seen by looking along the direction of travel, is always at the same orientation. We quickly showed that radio waves from pulsars are indeed polarized, and this led us immediately to a new way of exploring the interstellar medium: we found we could measure its magnetic field.

A magnetic compass on Earth is tracing a magnetic field which points (more or less) to the North Pole. This field has a strength of around 1 gauss;[24] the unit is named after Carl Friedrich Gauss (1777–1855). The field originates in electric currents in the dense molten iron core of the Earth. A magnetic field also exists in the interstellar medium. Here it is sustained by bulk movements of electrons making currents in the thin ionized gas; not surprisingly this can sustain only a weak field, about a millionth of the magnetic field on Earth. It would be very difficult to measure directly, even if we could go there, but fortunately we can do this remotely by radio measurements.

Any movement of an electron makes an electric current, and any electric current in a magnetic field feels a force proportional to the strength of the field. An electron oscillating in response to a radio wave is no exception, and the motion of an electron in response to the radio wave from a pulsar is affected by the magnetic field in interstellar space. The movement of the electron reacts back on the radio wave, and changes the direction of the oscillating electric field. The effect is to change the direction of polarization of the radio wave. The effect was first observed in light by Michael Faraday (1791–1867), and it is named Faraday rotation.

Faraday first demonstrated this effect in 1845, using a large magnetic field to rotate the polarization of light in a block of glass. In the interstellar medium the amount of rotation is small, being proportional both to the strength of the weak magnetic field and to the density of the thin population of electrons, but over distances of light-years it can add up to several complete rotations of the plane of polarization.

Furthermore, the amount of rotation depends on radio frequency, so measurements at several radio frequencies can be combined to find the actual amount of rotation.

One problem is that there is no Faraday rotation when the radio wave is travelling across the magnetic field rather than along it. In the first publication on the Faraday effect on pulsar radio waves, I described measurements on a pulsar which was being viewed, as we found out later, almost exactly across the field, giving a low and uncertain value for the strength of the field. The measurement is now routinely made on every known pulsar. Given the electron population of the line of sight, which is already found from the dispersion delay, the Faraday rotation gives an average strength of the magnetic field along the line of sight to each pulsar. A map of the magnetic field can then be constructed covering most of the Galaxy. The field runs along the spiral arms, and extends throughout the thick disc of ionized hydrogen.

Twinkle Twinkle Little Radio Star

Stars twinkle in the night sky because we see them through the Earth's atmosphere, which is never completely smooth. 'Little' star is appropriate; the planets appear as large discs, and twinkle very little or not at all. Light from a star may have travelled thousands of years without any effect from the interstellar medium, only to be diffracted and distorted by a turbulent atmosphere. No wonder that astronomers use telescopes mounted in spacecraft, high above the atmosphere, like the Hubble Space Telescope.

Radio waves are not much affected by our atmosphere, but in contrast to light waves they are seriously affected by the electrons of the interstellar medium. We have already seen how electrons in the line of sight can delay a radio pulse, and change its polarization; these effects

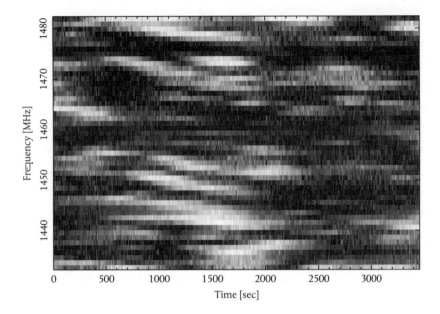

FIGURE 22 Varying radio signal from a scintillating pulsar, recorded simultaneously at a range of frequencies from 1440 to 1480 MHz over a period of an hour. The signal fades with a timescale of 20 minutes, with a frequency structure of 5 MHz. *Patrick Weltevrede, Jodrell Bank Observatory.*

tell us about the average smoothed out distribution of the electrons and the magnetic field. The radio waves we receive from small-diameter radio sources, and particularly the pulsars, twinkle like the visible stars but more slowly. We call this *scintillation,* and we use it to investigate the irregular nature of the ionized interstellar medium. Figure 22 is a recording of the variation in signal strength of a radio source 1000 light-years away. It happens to be a pulsar, but in this recording the individual pulses do not show up, only the signal strength averaged over some minutes. Figure 22 shows a recording made simultaneously at several adjacent radio frequencies, with obvious differences. (A similar scintillation in light would show changing colours.) From this type of recording we can find the scale and depth of the turbulence in the interstellar medium.

Radio scintillation also occurs in two, more local, circumstances where radio waves encounter free electrons: these are the terrestrial ionosphere and the solar corona. The ionosphere, lying high up at 100 kilometres or more in the Earth's atmosphere, is ionized mainly by sunlight. Long-wave radio, at radio frequencies below those used in astronomy, can be reflected by the ionosphere; long-distance, long-wave, radio communication relies on ionospheric reflection to overcome the curvature of the Earth. Radio astronomy sometimes has to take account of some bending of radio waves as they pass through the ionosphere, especially if observations are being made to find accurate positions of radio sources. The ionosphere is not a uniform layer, and there are sufficient irregularities in it to cause scintillation to be seen in astronomical observations at fairly low radio frequencies. A similar phenomenon is familiar to amateur radio enthusiasts in long-range, low-frequency radio using ionospheric reflection, when the signal fades and recovers on time scales of seconds or minutes. Ionospheric scintillation of the strong radio sources which had recently been discovered at Cambridge was followed through a year of observations by Antony Hewish. He found that the effect was greatest soon after midnight, which was surprising considering that solar radiation was supposed to be the origin of the ionization. This turned out to be related to a phenomenon observed in routine radar investigations of the ionosphere, known as 'spread F' echoes. The main region of the ionosphere is known as the F-region; this persists through the night even though it is generated by sunlight. Apparently the ionization in the F-region becomes more irregular around and soon after midnight; the cause of this is still unknown.

At the time of these investigations, around 1950, it was uncertain whether the whole of the fluctuations in signal strength observed in discrete radio sources were due to scintillation, or partly due to actual variations in the source itself. This was tested in a cooperative observation between the two UK radio observatories, Jodrell Bank and Cambridge.[25] Scintillation due to a discrete ionized region, such as the ionosphere, has a very familiar analogy in the refraction of sunlight in a swimming pool. The pattern of light on the bottom of the pool has the same scale as the ripples on the surface of the water, moving and changing with the ripples. Ionospheric scintillation has the same effect, making a pattern of signal intensity on the ground which is closely related to the scale of irregularities in the ionosphere. This pattern is only a few kilometres across, so recordings at the two observatories, 200 kilometres apart, should show fluctuations which are totally unrelated, unless the source itself was fluctuating. As expected, the test showed no correlation, and unsurprisingly the sources themselves were shown to be constant, at least on a short time scale.

The other, rather less local, cause of radio scintillation is in the solar corona, and in its extension to interplanetary space. Hewish, who had investigated ionospheric scintillation, turned his attention to the effect of the Sun on the recently identified strong radio source, the Crab Nebula. It happens that the Sun, in its annual movement across the sky, passes close to the Crab Nebula in June every year. Having recently discovered that refraction in the Earth's ionosphere could change the apparent position of a radio source, Ken Machin and I wrote a paper[26] suggesting that refraction might be seen in the outer corona of the Sun if we watched the position of the Crab Nebula day by day through June. No such luck; instead the radio source became a large blur due to random refraction and diffraction in the outer corona. This was the observation that suggested to Hewish that he should look at scintillation in a considerable number of radio sources distributed over the sky, to look at the whole of interplanetary space and possibly detect clouds of ionized hydrogen streaming out from the Sun. So was born the idea of a large telescope on the low frequency of 81 MHz (wavelength 3.7 metres) with which he and Jocelyn Bell serendipitously discovered an entirely different fluctuating radio source, the first pulsar (see Chapter 6).

4

Cosmic Rays, the Synchrotron, and Molecules

Cosmic Ray Air Showers

The Earth is continuously under bombardment by high-energy particles from outer space. These *cosmic rays* are mostly protons and electrons, the constituents of ordinary matter, travelling through interstellar space as individual particles with very high energy. They were discovered and named long before the discovery of cosmic radio waves.

When an energetic cosmic ray collides with an atom in the atmosphere, it explodes into several particles, which in turn collide and develop a cascade of billions of elementary particles. This is a cosmic ray air shower, comprising mainly protons and electrons but with some more exotic particles called positrons, muons, and pions. The tracks of ionization made by the particles in a cosmic ray shower can be made visible in a cloud chamber, where drops of water condense on to the tracks (Figure 23). Most cosmic ray particles are stopped high in the atmosphere. Their ionizing effect on the atmosphere therefore increases with height; this was demonstrated by Victor Hess (1883–1964) in a classic balloon flight in 1912. A charged electroscope becomes discharged if the air surrounding it contains free electrons.

Plate 92

FIGURE 23 Tracks of cosmic rays in cloud chamber. Collisions in lead sheets produce a shower of particles, like the larger scale cosmic ray showers in the atmosphere. *Proceedings of the Echo Lake Symposium.*

Hess took an electroscope in a hydrogen-filled balloon on a perilous flight to an altitude of 5 kilometres, and found that the ionization of the atmosphere increased tenfold. The significance of this observation was not appreciated until many years later; in 1937 he received a Nobel Prize for this work.

The Radio Observatory at Jodrell Bank originated from a proposal to detect individual cosmic ray showers by radar. Bernard Lovell (1913–2012) came to Manchester to work with Patrick Blackett (1897–1974), who succeeded William Bragg (1890–1971) as professor of physics in 1937. Blackett[27] was interested in some very-high-energy particles which remained in a shower even after multiple collisions in the atmosphere; his investigations of these particles in cloud chamber experiments at Cambridge earned him a Nobel Prize, awarded in 1948. On Blackett's arrival in Manchester, the direction of the physics department changed dramatically from crystallography to particle physics, and a new laboratory for high-energy physics was established. However, the new science was cut short in 1939 when both Lovell and Blackett were immersed in World War II, Blackett working on operational research and Lovell on radar. Lovell was involved in the rapid development of airborne radar, but early in the war he had visited one of the Chain Home coastal defence radars. Watching the radar screens he had seen echoes when no aircraft were anywhere near the radar station. He speculated that these might be echoes from the cloud of charged particles created in cosmic ray air showers, and he and Blackett wrote a paper on the possibility of using radar to investigate cosmic ray showers, and possibly to find the origin of the primary cosmic ray particles. Another pioneer of radio astronomy, James Hey, whom we have already encountered in Chapter 2, was already using wartime radar experience with an army portable radar to investigate cosmic radio noise and radar echoes from meteor trails, and this seemed a possible route to detecting echoes from air showers. When Lovell returned to Manchester after the war he borrowed a radar set from Hey and with his help set it up in Manchester to look for these echoes. His first attempt was in the courtyard of the main building of the Victoria University (as it then was), but the sensitive receiver of the radar was overwhelmed by electrical interference, especially from street trams, and he had to move out to a university field station at Jodrell Bank, where he was supposed to be the short-term guest of the Professor of Botany. He never left, and Botany eventually moved out and left him in sole occupation. Lovell and his colleagues immediately found some very interesting echoes from trails left by meteors high in the atmosphere, but none from air showers.

Lovell decided he needed a bigger receiving antenna if he was to find radar echoes from cosmic ray showers, and so started the sequence of large radio telescopes which made the Jodrell Bank Observatory a leader in radio astronomy. The echoes were never found, and the telescopes led to research in entirely different directions. But the story does not end there. In 1952 Galbraith and Jelley, working at the Atomic Energy Research Establishment, showed that the electrons in the shower radiate a detectable flash of light.[28] The electrons are travelling at nearly the velocity of light through the atmosphere, in which light and radio travel appreciably more slowly than the free space velocity of light and slower than the shower of electrons. Under these conditions they must radiate *Cerenkov radiation*. This type of radiation was first characterized by Pavel Cerenkov (1904–1990) to explain the blue glow of light emitted by a beam of high-energy particles made in a laboratory nuclear reactor when they passed through a surrounding water tank; here again there were charged particles moving faster than the velocity of light in water, which is less than the velocity in free space. (In 1958 Cerenkov became another of the many Nobel Prize winners associated one way and another with radio astronomy). John Jelley pointed out that there should also be a pulse of radio, which should be detectable. He came to Jodrell Bank in 1964, bringing particle detectors which would respond to a cosmic ray shower, while we provided a sensitive receiver and a small upward looking radio telescope. The particle detector was arranged to trigger a display of the telescope output on a cathode ray tube. We did indeed detect the radio pulse from several showers, and we were the first to do so.[29] Several cosmic ray observatories are now using similar radio techniques to characterize air showers, but Jodrell Bank Observatory itself has now left the subject entirely. The outcome of any scientific research is indeed unpredictable!

The use of Cerenkov radiation in detecting cosmic ray air showers has been developed in a major international observatory (Figure 24), named after Pierre Auger (1899–1993), who discovered extensive air showers. In this observatory, the Cerenkov radiation is generated as flashes of light when particles energetic enough to reach the ground pass through sealed tanks of water particles. The tanks are separated by 1.5 kilometres; 1600 of them are spread over a wide area, so that any shower falling within an area of several thousand square kilometres will be detected by light flashes in several adjacent locations. There are also radio pulse detectors, and a network of light detectors which can pick up the flash of fluorescent light caused by the ionization of nitrogen molecules high in the atmosphere. Cosmic ray showers have energies mostly in the range 10^9–10^{16} eV (electron volts); this is many orders of magnitude greater than the energy of a photon of light, which is of the order of 1 eV. The Pierre Auger Observatory is dedicated to detecting cosmic rays with the very highest energies, in the range 10^{18}–10^{20} eV. These are rare events; a cosmic ray with energy 10^{19} eV arrives on Earth at the low rate of 1 per square kilometre per year.

Air showers are also generated by energetic gamma-rays, which are photons at the opposite end of the electromagnetic spectrum from

FIGURE 24 Part of the Pierre Auger Cosmic Ray Observatory. Shower particles are detected by their Cerenkov radiation in tanks of water, and ultraviolet light excited by the shower is detected. There are 1600 of these detectors, spread over an area of more than 3000 km². *Pierre Auger Collaboration.*

radio. These gamma-ray showers may be distinguished from those generated by particles if the shape of the shower can be observed as it develops in the atmosphere. This has been shown to be possible in two new observatories, known as HESS (High Energy Stereoscopic System) in Namibia and MAGIC[30] at the Instituto Astrofisica de Canarias on the island of La Palma. These essentially photograph the shower stereoscopically and in sufficient detail to distinguish the two types of shower. Gamma-rays are particularly interesting in research on the sources of high-energy radiation, since they travel directly from the source to the observer without deviation by the large-scale magnetic field of the Galaxy.

Most cosmic ray particles are believed to derive their energy from acceleration processes in the Milky Way, involving strong shock waves produced in supernova explosions. This process can account for the acceleration of protons to about 10^{16} eV, but the acceleration mechanism for the very highest energy cosmic rays is not understood. A difficulty in finding their origin is that charged particles do not travel through the Galaxy in a straight line. The Galaxy is pervaded by a magnetic field, which is weak by terrestrial standards but strong enough to deviate charged particles and confuse their origin. The observatories are able to distinguish the cosmic ray showers from the gamma-ray showers, and have found that pulsars are the origin of many of the gamma-rays. Whether the high-energy particles of cosmic rays are also generated by pulsars is an open question.

Synchrotron Radiation

We now return to the astronomy of our Galaxy, and make the connection between cosmic rays and the background radio waves discovered by Jansky and Reber. The connection is via the machines used in high-energy physics. Laboratory particle accelerators can make particles like cosmic rays, although they cannot reach anywhere near the highest energies. High-energy electrons and protons are routinely produced in synchrotron accelerators in research laboratories such as CERN. In these machines electrons race round a circular track, picking up energy at a series of accelerating stations and reaching almost to the velocity of light. A long series of larger and larger machines have been built, to produce particles with higher and higher energies. In each machine the energy limit is reached when the particles are losing energy by a fundamental radiation process as fast as they are being accelerated. This process of energy loss is *synchrotron radiation*, which is radiation produced by any charged particle forced by a magnetic field to travel along a curved path. It was named for its effect in synchrotron machines, and it limits the energy that can be reached in the machine. The tighter the curvature, the more does the particle radiate; the most energetic machines have to be designed with tracks with the lowest possible curvature, so the beams have to be confined to circular paths with radii measured in kilometres. The theory of this radiation was first worked out by Julian Schwinger (1918–1994), and for a time it was named Schwinger radiation. It is also named magnetobremsstrahlung, which means magnetic braking radiation. Synchrotron radiation can be very useful, and some machines are designed specifically to produce intense beams of electromagnetic radiation, which are used as sources of powerful X-rays and ultraviolet light.

It was a Swedish physicist, Hannes Alfvén (1908–1995),[31] (Figure 25) who first found the connection between the laboratory synchrotron radiation and cosmic radio waves. Working with a colleague, Nicolai Herlofson (1916–2004), who had recently spent three years in England both with Ryle's group in Cambridge and with Lovell at Jodrell Bank, he proposed that the galactic radio waves were generated by cosmic rays generated by stars and moving in a magnetic field close to the stars. Then a German astrophysicist, Karl-Otto Kiepenheuer (1910–1975), while visiting Yerkes observatory in America, correctly extended the idea to cosmic rays pervading the whole of the Galaxy. Throughout the whole of our Galaxy, the Milky Way, there is a magnetic field which forces all cosmic ray particles to follow a curved path, as they do in a synchrotron. The radiation that they produce as they do so is the same synchrotron radiation, but the curvature of their tracks is so slight that their radiation is at long radio wavelengths, and not at the short wavelengths of light and X-rays. The origin of the intense cosmic radio waves discovered by Karl Jansky was solved. The full analysis of cosmic synchrotron radiation was done in Russia, by Vitaly Ginzburg (1916–2009) of the P.N. Lebedev Institute in Moscow, completing a remarkable international trail of discovery and analysis.

FIGURE 25 Hannes Alfvén, with a diagram of synchrotron radiation, which he correctly identified as the source of radio emission from the Milky Way.

The strength and spectrum of synchrotron radiation depends on the energy of the particles and the curvature of their tracks. In a laboratory machine designed to produce light and X-rays the tracks are tightly curved, with a radius of some tens of metres. In interstellar space the tracks are curved by a very weak magnetic field; consequently the radius of their curved tracks is comparable to the distance between the stars. Synchrotron radiation from cosmic rays is therefore only important at radio wavelengths, and it has a steep spectrum which falls off rapidly towards centimetre and millimetre wavelengths. The spectrum is also related to the energy distribution in the cosmic ray electrons, which again emphasizes the radio emission at long wavelengths.

Another characteristic of synchrotron radiation which is often observed in many types of radio source, and especially galactic radiation, is polarization.[32] When the track of an electron is curved by a magnetic field, the electric field which it radiates is aligned in the plane in which it is moving. If, as in a particle accelerator, the track of the electron is perpendicular to the magnetic field, the electric field of the radiation is exactly defined and the radio emission in the plane of the curved track is fully linearly polarized. There is an angular spread of radiation; on either side of the plane the radiation is polarized, but it has a component of circular polarization, with opposite hands on either side. Averaging over the whole beam, the opposite hands of circular cancel out, leaving an average high degree of linear polarization, which may typically be as high as 70%. Even if the electron tracks are randomly oriented, the radiation from a cloud of electrons will have a preferred component perpendicular to the magnetic field, giving an easily observable degree of linear polarization. The technique for observing this may be visualized by considering the effect of rotating the dipole at the focus of a television satellite receiver; the radio signal is usually linearly polarized to some degree, and the signal varies according to the alignment of the dipole and the radiation.

It was the polarization of synchrotron radiation that finally proved its relevance in astronomy. The Soviet astrophysicist Iosef Shklovsky (1916–1985) suggested in 1952 that the light from the supernova remnant known as the Crab Nebula (see Chapter 6) might be synchrotron radiation, which might show linear polarization. In 1954 Viktor Dombrovski (1914–1972) and Mikhail Vashakidsze (1909–1956) used a polaroid filter on a photometer measuring the light from the Crab Nebula (polaroid is used by automobile drivers and fishermen to cut out polarized light reflected from a water surface). The intensity of the light was different at different angles of the polaroid, proving that the light is polarized and must therefore be synchrotron light. Professor Jan Oort of Leiden, who appeared in Chapter 3 as the father of the radio hydrogen line, and his colleague Theodore Walraven (1916–2008) then took photographs of the Nebula using a polaroid filter at telescopes on the Pic du Midi in France and the 200-inch (508-centimetre) Palomar telescope in California, showing a complex structure of very highly polarized emission. The same effect was soon demonstrated in the radio from the Galaxy, confirming the theory that the intense radiation discovered by Jansky is synchrotron.

Synchrotron radiation at radio wavelengths occurs in many kinds of radio source, from within the Solar System to the furthermost galaxies. Closest to home, it has been observed and mapped by the Cassini spacecraft in the atmosphere of Jupiter (see Chapter 2). This huge planet has a stronger magnetic field than our Earth, and its atmosphere is filled with particles with energies in the low cosmic ray energy range. Figure 14 shows that these radiating particles are circulating high in Jupiter's atmosphere; the planet is apparently acting as a synchrotron accelerator millions of times larger than our terrestrial machines, but still tiny compared with the grand scale of the Milky Way.

Molecules in the Galaxy

Synchrotron radiation, like white light, is spread over a wide range of frequencies, while the 21-centimetre hydrogen line radiation, which has been so useful in delineating the structure of our Galaxy, is concentrated in a spectral line, like the light from a sodium street lamp. Another spectral line, at a wavelength of 18 centimetres, was discovered serendipitously some time after the dramatic discovery of the hydrogen line, and eventually attributed to the hydroxyl radical OH. In this molecule, as in the hydrogen atom, there is a split in the ground energy level due to electron spin, but in addition each of the split levels is split again due to the spin of the whole molecule. The OH spectral line is split into four, at 1612, 1665, 1667, and 1720 MHz. These lines are often very bright (in radio terms), because they are amplified by a 'maser' action. This is Microwave Amplification by Stimulated Emission of Radiation, and is the radio equivalent of the laser for light. Bright OH spectral lines can be observed in the atmospheres of stars, adding to the possibilities of measuring the rotation curve in the outer parts of the Galaxy.

The dark patches which can easily be seen by the naked eye in the Milky Way are regions where distant stars are obscured by huge clouds of dust. The dust grains are around 1 micron (a thousandth of a millimetre) in size, and are probably made of loose aggregations of carbon and other common elements. These are the nurseries of our Galaxy, where new stars are born. Within the clouds are many and various molecules which can be observed by radio, particularly at shorter wavelengths in the centimetre and millimetre range (Figure 26). Among these are familiar substances such as water, alcohol (both methyl and ethyl), ether, ammonia, acetylene, hydrogen cyanide, and formaldehyde. These molecules have more complicated sets of energy levels than mono-atomic hydrogen, and the identification of the radio spectral lines they produce

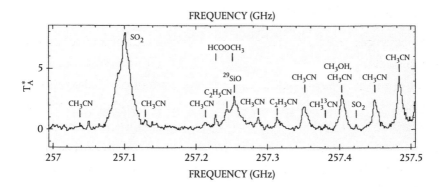

FIGURE 26 A spectrum of radio emission from a molecular cloud, showing many spectral lines generated by molecules. (*Blake et al.* 1986). *Reproduced by permission of the AAS.*

Table 1 Molecules detected in interstellar clouds

No. of. Atoms	Species
2	A_1F...A_1C_1...$Al.O$...C_2...CH...CH^+...CN...CN^+... CO...CO^+...CF^+...$CPvCS$...HD...HC_1...HCl^+... HF...KC_1...NH....N_2...NO...NO^+...NS...$NaCl$... OH...PN...SH...SO...SO^+...SiC...SiN...SiO...SiS
3	$AlNC$...$AlOH$...C_3...C_2H...C_2o...C_2S...C_2P...HCN... $FeCN$...H_3^+...H_2C...H_2Cl^+...H_2O...H_2O^+...H_2D^+... HD_2^+...HDO...D_2O...H_2S...HCN...HNC...HCO... DCO...HCO^+...HCP...HCS^+...HNC...DNC...HN_2^+... HNO...HOC^+...KCN...$MgCN$...$MgNC$... NH_2...N_2H^+...N_2D^+...N_2O...$NaCN$...$NaOH$... OCS...O_3...SO_2...$c\text{-}SiC2$...$SiCN$...$SiNC$
4	CH_3...$c\text{-}C_3H$...$1\text{-}C_3H$...C_3N...C_3N...C_3O...C_3S... C_2H_2...H_2CN...H_2CN^+...$HCNH^+$...H_2CO... $HDCO$...D_2CO...H_2CS....$HCCN$...$HNCO$... $HNCS$...$HOCO^+$...NH_3...NH_2D...NHD_2...ND_3...SiC_3...C_4
5	C_5...C_4H...C_4Si...C_5N...$1C_3H_2$...cC_3H_2...CH_2CN... CH_4...HC_3N....HC_2NC....$HCOOH$....H_2CHN... H_2C_2O...H_2NCN...HNC_3...H_2COH^+...SiH_4
6	C_5H...C_5N...C_2H_4...CH_3CN...CH_3NC... CH_3OH...CH_3SH...HC_3NH^+...$1\text{-}H_2C_4$... HC_2CHO...NH_2CHO...HC_4N
7	C_6H...CH_2CHCN...CH_3C_2H...HC_5N... $HCOCH_3$...NH_2CH_3...$c\text{-}C_2H_4O$...CH_2CHOH
8	CH_3C_3N...$HCOOCH_3$...$CH3COOH$...C_7H... H_2C_6...CH_2OHCHO...CH_2CHCHO... CH_2CCHCN...NH_2CH_2CN
9	CH_3C_4H...CH_3CH_2CN...$(CH_3)_2O$...CH_3CH_2OH... HC_7N...C_8H...CH_3CNH_2...CH_3CHCH_2
10	CH_3C_5N...$(CH_3)_2CO$...$(CH_2OH)_2$...CH_3CH_2CHO
11	HC_9N...C_2H_5OCHO...CH_3C_6H...$HCOC_2H_5$
12	C_6H_6...C_3H_7CN
13	$HC_{11}N$

has become a new science of astrochemistry, in which astronomers, spectroscopists, and laboratory chemists join forces. Some remarkably complex molecules have been shown to exist in the clouds, which are known as the Giant Molecular Clouds. Harry Kroto (b. 1939) and his colleagues at the University of Sussex in the UK investigated the spectrum of the linear chain molecule HC_5N in laboratory measurements, and suggested it might be found in the clouds. Indeed it was, in large quantities, suggesting a search for larger molecules such as HC_7N, which was duly found, and eventually $HC_{11}N$, the longest chain molecule found so far. The list of molecules identified in radio and infrared from the molecular clouds now runs to over 100 entries (Table 1). Millimetre wave astronomy, both for spectral lines and for continuum radiation, is now so important that a very large radio telescope, ALMA, is devoted to it (see Chapter 11).

The story of the long chain molecule HC_5N and its longer brothers had a remarkable sequel. Kroto collaborated with chemists who had techniques for synthesizing carbon compounds using energetic particle beams and measuring the molecular weights of the compounds that emerged. To their surprise, a molecule with a mass of 60 carbon atoms turned up frequently in the analysis. They had discovered C_{60}, a molecule with a spherical arrangement of 60 carbon atoms, which is now known as buckminsterfullerene. This discovery of 'buckyballs' led to further discoveries of carbon configurations, now familiar as nanotubes and graphene, the film with mono-atomic thickness.

Within a degree or two of the centre of the Milky Way galaxy there is a Central Molecular Zone, with a complex structure which can be sorted out from measurements of the Doppler shifts of another important spectral line, from carbon monoxide, CO. This molecule, which is common in the Galaxy, has a prominent radio spectral line at a wavelength of 2.6 millimetres.

The complexity of the radio spectrum of a typical molecule is related to the way in which energy is stored in a set of energy levels related to rotations of the whole molecule and vibrations within it. Transitions between vibrational energy levels have typical energies of 0.1 to 0.01 electron volts, giving rise to spectral lines in infrared light. Transitions between rotational states give energy steps typically of 0.001 electron volts, corresponding to wavelengths in the centimetre or millimetre range. The CO molecule has a comparatively simple radio spectrum; it is the angular momentum of the rotation which is constrained to discrete levels, which are labelled $J = 0,1,2$, etc., and a transition between $J = 0$ and $J = 1$ emits a radio wave at 115 GHz. Larger rotational energies are possible, even in the cold interstellar gas, but the transitions have a higher energy and the spectral lines are outside the highest radio frequency range. Remarkably, some of these higher energy transitions can be observed in very distant galaxies which are receding at high velocities, when the redshift of their whole spectrum brings the line frequencies down to the observable band.

It is surprising that atoms can get together and combine to form molecules like CO, when the density is so low and the chances of collisions between atoms such as carbon and oxygen should be negligible. This is where the dust grains come into the story. Collisions with such comparatively large objects are frequent. Atoms stick to the surface of dust grains, and can migrate across the surface to encounter their future partners. Molecular hydrogen, H_2, is probably formed this way. It has no spectral lines in radio, but spectroscopy in the visual detects H_2 as an absorption feature in the light from distant stars, and it is known to be widespread and associated closely with CO. Radio surveys address the interstellar medium directly, and are easier to use than stellar spectra, so surveys of the CO 2.6 millimetre line are used as a proxy for H_2. The CO line is so prominent that it raises its own difficulty: there is so much of it that it is impossible to see through thick clouds of CO, and maps made close to the galactic plane cannot show the more distant parts. There is too much carbon monoxide! Fortunately carbon exists in several isotopes, with extra neutrons increasing the mass and changing the frequency of the CO spectral line by a small fraction. The common species of carbon is ^{12}C, but the less common species ^{13}C, ^{17}C, and ^{18}C have been observed in galactic carbon monoxide. These species are much rarer than ^{12}C, and surveys of carbon monoxide with the more massive isotopes are commonly used to map distant gas clouds, avoiding the problem of overfilled lines of sight. Figure 27 is an example. This shows the structure close to the centre of the Galaxy, where the dynamics depart dramatically from the smooth curve of rotation shown in Figure 17.

The discovery of the radio spectral lines from H, OH, and CO aroused interest in the new subject of astrochemistry; many more lines were predicted and observed at shorter wavelengths in the molecular clouds, particularly in the Orion Nebula and in a cloud known as Sagittarius A, near the centre of the Galaxy. Many of these spectral lines are attributed to molecules of considerable complexity, some of which are organic molecules which might be building blocks in the synthesis of life. The list includes 167 species of molecules (including some which are ionized, indicated by the + superscript) with numbers of atoms up to 13.

The list is even longer if isotopic species are included (some deuterated molecules, with deuterium, D, in place of hydrogen, H, are included in the list). The record number of atoms in the radio list has been overtaken by the detection in infrared of buckminsterfullerene C_{60} and the next fullerene C_{70}; these have been found in the denser environment of a cloud round a star, known as a planetary nebula.

The pattern of spectral lines becomes very complex for these multiatomic species. Even water H_2O has complex rotational energy levels,

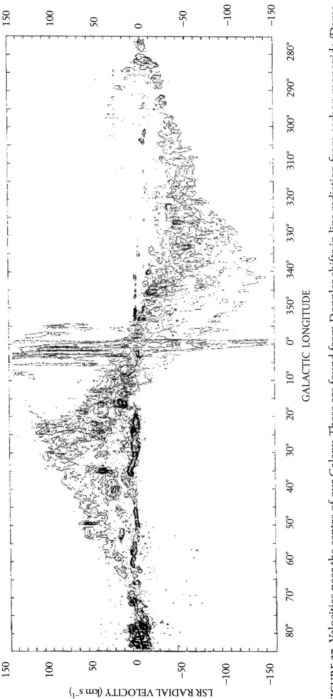

FIGURE 27 Velocities near the centre of our Galaxy. These are found from Doppler shifts in line radiation from carbon monoxide (Dame et al. 1987). *Reproduced by permission of the AAS.*

because it is a molecule with no single axis of symmetry, like OH or CO for example. The atoms in a water molecule form a bent line, like a boomerang, and rotation about more than one axis is subject to separate quantum rules. Ammonia, NH_3, appears in the Table; here the energy levels are related to a peculiarity of structure, in which the three hydrogen atoms form a triangle and the nitrogen atom pops from one side to the other like the lid of a tin can, an oscillation known as the oil-can mode. The energy of the oscillation, as always, is subject to the rules of quantum mechanics.

Many of us are delighted to see ethyl alcohol, C_2H_5OH, in the list, especially as it is found in vast quantities in a structure known as the galactic bar. Thoughts of scooping it up with a space probe and bringing it back to Earth can be dismissed; the volume of the clouds containing it is also vast, and it could only be collected molecule by molecule rather than scooped up in some sort of cosmic funnel.

5

Radio Galaxies and Quasars

Radio Stars or Radio Galaxies?

The cosmic radio waves discovered by Jansky and Reber came from our galaxy, the Milky Way. But did they come from diffuse clouds of gas, or from stars? Without radio telescopes with narrow beams there seemed no way to find out. As often happens, an answer came almost by accident. In 1946 James Hey, while mapping the radio emission from the Milky Way, noticed a remarkable fluctuating radio signal from somewhere in the constellation of Cygnus, which showed that there was a star-like radio source somewhere in the broad beam of his radio telescope (Figure 28). Hey and others already knew that the Sun was a powerful source of radio, and it was natural to suggest that the whole of the galactic radio might come from stars. It turned out, as we have seen in Chapter 3, that this was not the answer to the question of the origin of the radio emission from the Milky Way, which came not from stars but from the space between them, the interstellar medium. Hey's discrete source was not even any kind of star, but instead the first observation of a radio source entirely outside our galaxy. This became known as the radio galaxy Cygnus A.

'Twinkle twinkle little star, How I wonder what you are' (see Chapter 3) applies as much to Hey's discovery as it does to the

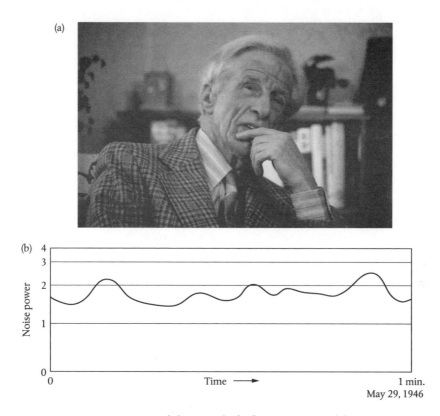

FIGURE 28 James Hey (a) and the record of a fluctuating signal from Cygnus (b). *(a) Hencoup/Galaxy. (b) Reprinted by permission from Macmillan Publishers Ltd: Nature Hey, J.S. et al 1946 'Fluctuations in Cosmic Radiation at Radio-Frequencies' vol. 158 © 1946.*

stars we see every clear night. It is not the stars themselves that twinkle, or *scintillate*, but turbulence in the Earth's atmosphere affecting light from the stars. Hey found the radio equivalent, a radio signal from a point-like object fluctuating as it passed through the ionized gas of the upper atmosphere. How we all wondered what this twinkling object might be! If its exact position could be found, then it might show up on photographs as some unusual kind of star.

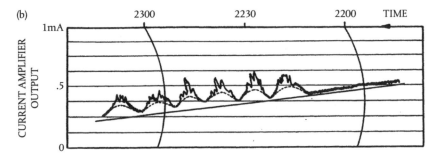

FIGURE 29 John Bolton (a) and the first interferometer record of Cygnus A (b). His interferometer used a single antenna, mounted on a cliff (Dover Heights) near Sydney. Pointing near the horizon, the antenna picked up radio waves reflected in the sea, giving the effect of an interferometer pair spaced by twice the cliff height. Cygnus A rose above the horizon at 2215 (time increases right to left). *(a) Courtesy of the Archives, California Institute of Technology. (b) CSIRO Radio Astronomy Image Archive.*

John Bolton[33] in Australia, and Martin Ryle in Cambridge, UK, both rose to the challenge. To pick out the discrete radio sources they used interferometers, pairs of antennas spaced by some hundreds of metres and connected to a single receiver. (Bolton's interferometer, Figure 29, actually consisted of a single antenna mounted on a cliff-top, over-looking the sea. The second antenna of his interferometer pair was the reflection of the antenna in the sea.) These interferometers had first been used for observing the Sun, and they were capable of distin-guishing between radio waves from sunspots and the much larger solar corona (see Chapter 2). As any compact source, such as a sun-spot, crosses the combined interferometer beam, its signal follows a sine wave pattern which distinguishes it from any wider source. Bolton's record of Cygnus A, made in 1948, shows this sinusoidal pat-tern, which comes from the small source, standing out from the base-line of the wider Milky Way signal. Cygnus A again showed the random and more rapid fluctuations already reported by Hey. The interference pattern also gave Bolton a position accurate to a fraction of a degree, although there was no obvious unusual star or any other visible object at that position. Soon he had found three more of these mysterious radio sources, and for those three he found reasonably accurate positions. It was now worth while to look at conventional sky photographs to see if there were any unusual objects which might be the radio sources. He did indeed find that all three were close to some very unusual and interesting visible objects, the extragalactic nebulae M87 and Centaurus A, and the Crab Nebula. The radio bright-est object, Cygnus A, remained unidentified.

Meanwhile Martin Ryle and others in Cambridge (including myself) were also using an interferometer system to study the Sun and its sun-spots, and were able to turn their attention to Cygnus A. All that was needed was to direct the interferometer pair of antennas higher in the sky, and wait for 24 hours while the rotation of the Earth swept the antenna beam across the cosmos and everything it might contain. Our first record, in 1948, showed Cygnus A beautifully, but to our sur-prise another interference pattern appeared on the record a few hours later. We built a larger antenna system, and in 1949 obtained the record in Figure 30. As can be seen, the second pattern was wider than the Cygnus A pattern, showing that the new source moved more slowly across the sky; this meant that it was closer to the North Pole. The time of the centre of the pattern gave the longitude (Right Ascension)

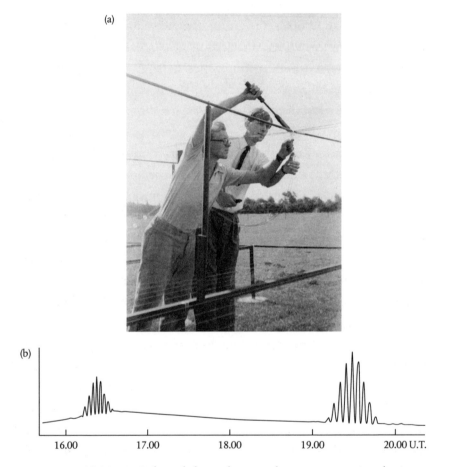

FIGURE 30 (a) Martin Ryle and the author are shown constructing the interferometer antenna. (b) The radio sources Cygnus A and Cassiopeia A recorded with an interferometer at Cambridge in 1949. *(a) Bruce Elsmore. b) Courtesy of Oxford University Press and the Monthly Notices of the Royal Astronomical Society.*

and the rate of angular movement gave the latitude (Declination). The new source was located in the constellation of Cassiopeia, and it later became known as Cas A. There was no obvious visible counterpart. But as more accurate positions were found, and with the help of astronomers with large telescopes, all five sources were identified. Two, in the constellations of Cassiopeia and in Taurus, were the visible remains of supernova explosions: they were indeed in our galaxy. But the others were extragalactic, completely outside our galaxy; they were galaxies which were much more powerful radio emitters than the whole of the Milky Way. And one of them, the original Cygnus A, was among the most distant objects known at that time.

Tracking down the exact position of Cygnus A needed a new, larger and more precise interferometer. This was built at Cambridge, using two 8-metre-diameter antennas originally used in Germany for a World War II radar system known as Würzburg. Nominally aligned on an East–West line, and set (by hand) at any angle above the horizon, they could observe any object as it crossed the Cambridge meridian. The receiver, built before the age of transistors and digitization, was very simple, and the recordings were ink lines on paper charts. Calibration was an interesting exercise for me as a PhD student; having first learnt the essentials of positional astronomy, I had to find the actual location of the two antennas and make corrections for the difference between their alignment and a precise East–West line. An accurate clock was borrowed from the Royal Greenwich Observatory. Observations were continued throughout a whole year in case the sources might be moving, either within themselves or because of variable refraction in the terrestrial ionosphere. This *transit interferometer* gave positions to an accuracy of 1 minute of arc; this is very rough by modern standards, but it was good enough to confirm the Australian identifications of two sources, one of which was the Crab Nebula, within our galaxy, and the other an extragalactic nebula in Virgo.

Finding the visible counterparts of Cas A and Cygnus A needed help from experienced optical astronomers.

David Dewhirst (1926–2012), of the Cambridge Observatory, was the first to see faint traces of both sources on a photographic plate. But it took the resources of the Mount Wilson and Palomar Observatories in California to find their true nature. Cas A turned out to be the remnant of a supernova explosion which had not been previously recorded, supporting the idea that the discrete sources were mostly within the Milky Way galaxy. The identification of Cyg A was a dramatic breakthrough; it was a most unusual extragalactic nebula at an astonishingly great distance. It was this discovery that changed radio astronomy from a local astrophysical interest to a major contributor to cosmology, which we explore in Chapters 8 and 9.

Cygnus A was the most elusive of the early radio sources. In 1951 there was only a suspicion that a small smudgy object on a survey photograph might be the faint visible image of this very bright radio source. So our accurate position was sent to California, where two remarkable and adventurous astronomers, Walter Baade (1893–1960) and Rudolf Minkowski (1895–1976), had access to the world's largest telescopes, 100 and 200 inches (508 centimetres) in diameter. The result was astounding. The little smudge turned out to be a faint galaxy which by chance was almost hidden behind a crowded field of stars in our own galaxy. The spectrum of light from this galaxy was amazing: it showed several familiar spectral lines, but the whole pattern was shifted towards longer wavelengths. This was the famous 'redshift', which showed that the galaxy was moving away from us at about 6% of the velocity of light (astronomers refer to this as a redshift $z = 0.06$). According to Hubble's law this meant it was at a very great distance, almost 10 million light-years away. The width of the spectral lines also showed very large internal velocities of up to 400 kilometres per second. Cygnus A, almost the brightest radio source in the whole sky, was one of the most distant and most unusual galaxies ever observed. This was surely the moment when radio astronomy showed its potential in observational cosmology.

It seemed incredible that such a distant galaxy could be such a strong radio emitter. Maybe the visible galaxy was only part of a much larger object; Baade and Minkowski therefore asked if the radio observers could actually measure the size of Cygnus A. Yes, indeed we could, using the new interferometer technique. All that was needed was an interferometer in which the spacing between the two antennas could be increased, making a progressively finer interference pattern. When the angular width of this pattern became smaller than the width of the source, the records would show a reduced interference pattern.

The measurement of the angular diameter of Cygnus A was done almost simultaneously in Sydney, Australia, and in England both at Cambridge and at Jodrell Bank. It was anyone's guess how far apart the two antennas should be, but at Sydney and Jodrell Bank the observers guessed they would have to be so far apart that they could only be joined together by a radio link. This took some time to organize; meanwhile at Cambridge there was the transit interferometer, whose two antennas were linked by cable but might not be not far enough apart. By good fortune this smaller spacing was about right, as Cygnus A turned out to be far larger than anyone had thought possible. The visible galaxy was several seconds of arc across, but the radio galaxy was found to be several minutes of arc across, about one-tenth of the width of the full Moon and more than 50 times larger than the faint smudge on the photographic plate. Soon afterwards the interferometer at Jodrell Bank gave the first indication of structure within the radio source. Mapping that structure in detail became a challenge which led eventually to the generation of the synthesis radio telescopes described in Chapter 10.

The Discovery of Quasars

Discovering more of these extragalactic radio sources needed radio telescopes with more sensitivity and better angular resolution. Larger antennas were needed, but single reflectors do not have good enough angular resolution. The first surveys, made at Cambridge in the 1950s, used pairs of large cylindrical reflectors. The third survey resulted in the 3C catalogue of 470 sources, which became the standard list of bright radio sources. The positions in this catalogue were accurate enough to stimulate an intensive search for optical counterparts, which eventually led to identifications of every source in the catalogue. At the same time the radio interferometer at Jodrell Bank was extended in an attempt to measure the diameters, and possibly the shapes, of these new radio sources.

The interferometer at Jodrell Bank was, from the start, designed to work with large separations between the pair of antennas, which were connected together by using a radio link. One of the pair was the 250-foot (76.2-metre) telescope now known as the Lovell Telescope; the other was a small lightweight dish mounted on a trailer. The trailer, with its dish and radio link, was taken to stations up to 115 kilometres from Jodrell Bank, giving interferometer spacings of up to 61,000 wavelengths. It should be possible at this spacing to measure the angular diameter of any source which was greater than about 3 seconds of arc. Most of the new radio sources were seen to be resolved into two components, rather like Cygnus A but apparently much smaller as it would appear if it were at a greater distance. But four radio sources remained unresolved; these evidently had a diameter no larger than the image of a star on most optical photographs. A further attempt was made using a shorter wavelength, so as to increase the number of wavelengths in the interferometer pair, and the observations ended with the conclusion that the diameters of these four objects were all less than 0.5 arcseconds. They became known as quasi-stellar radio sources, a name soon shortened to *quasars*. Their nature was a mystery; they could be stars in our galaxy, or they could be at great distances. In either case they were prodigiously powerful.

The search for optical identifications was mainly undertaken by Allan Sandage (1926–2010), Maarten Schmidt (b. 1929), and their colleagues at Mount Palomar and Mount Wilson Observatories. The results were a revelation. At one stage, around 1960, the possibility of local radio stars was resurrected, but instead the prospect of cosmological exploration to previously inaccessible depths was opened. Three of the 470 catalogued 3C radio sources whose radio diameters were already known to be less than an arcsecond were tentatively identified with star-like objects. No distance could be found, and they might still be identified as stars within our Galaxy, apart from their strange spectra. There was also another, 3C 273, which was apparently at the position of a 13th magnitude star, bright enough to be seen with a modest sized amateur telescope. Such an improbable identification needed verification, and a more precise position was needed. A novel method of position-finding was brought into play for this vital measurement, using *lunar occultation*.

The track of the Moon across the sky, as seen from any part of the world, is known and precisely predictable. If it crosses the position of a radio source, the signal will be cut off and reappear at times which give the position of the source. The method was tried out successfully on another catalogued 3C source at Jodrell Bank by Cyril Hazard (b. 1928) in 1960, but unfortunately the track of the Moon across the sky, as seen from England, did not cross the position of the 3C 273 quasar. However, it was predicted that as seen from Australia, the track of the Moon would cross the position of the quasar twice in 1962, so that it could be observed by using the 210-foot Parkes radio telescope. Unfortunately, the Moon was very low in the sky at the critical moment, and the telescope mounting had to be modified to allow it to observe so near the horizon. Nevertheless, Hazard succeeded in timing the moments when the radio source disappeared and reappeared. Knowing the movement of the Moon, these times gave a position accurate to about 1 second of arc, showing that the star-like object was indeed the radio source. This stimulated redoubled efforts to interpret the optical spectrum, in which the spectral lines were all in the wrong place. All became clear when it was realized that they were all shifted towards the red end of the spectrum, meaning that the source was moving away at no less than one-sixth of the speed of light. In astrophysical terms, it had the unprecedentedly large redshift $z = 0.158$. The other three suspected star-like objects were interpreted soon afterwards, showing even larger redshifts of 0.3675, 0.425, and 0.545. These were an entirely new class of objects, which deserved their new name, the quasars. Soon afterwards, in 1965, Maarten Schmidt found another quasar, 3C9, with a redshift $z = 2.012$; in the spectrum of this quasar the lines appeared at wavelengths three times their normal positions.

The quasars were not the only identifications. Bolton had already suggested that one of his four original radio sources was an unusual galaxy in the constellation of Centaurus. This was followed by several other galaxies, and a category of radio galaxies was established. These were not at the phenomenally large distances of the quasars, so the question arose: were these radio galaxies distinct from the quasars? The link between the two categories actually dates back to 1918, when Heber Curtis (1872–1942) published a survey of nebulae, even before some of these were known to be extragalactic. He noted that the galaxy M87 has an unusual jet-like protrusion (see Plate 8). In 1956 Walter Baade found that the light from this jet was optically polarized: we now know that this is a sure sign that it is synchrotron radiation.

A similar jet was then found in the quasar 3C 273, and the radio galaxy and the quasar were linked by the suggestion that in both cases the jet originated in an unusually active central nucleus in a more or less conspicuous galaxy. There was already such a category of galaxies, which were assembled from a survey of optical spectra in 1940 by Carl Seyfert (1911–1960). These Seyfert galaxies evidently had Active Galactic Nuclei, or AGNs. Finally Alan Sandage in 1965 discovered a large number of visible quasi-stellar objects with the same characteristics as 3C 273, but which did not appear in any radio catalogues. There were some hundreds more of these radio-quiet quasars than the radio-loud ones known at the time. It was time to put all these discoveries together in a unified theory.

Black Holes

The key to the whole display of active galactic nuclei, quasars, and radio galaxies was proposed in 1964 independently by Edwin Salpeter (1924–2008) in the USA and Yakov Zel'dovich in the USSR. They pointed out that the AGNs required a source of energy which was so large that it could only be accounted for by the accretion of large quantities of matter on to a black hole. We have already seen in Chapter 3 that our galaxy, the Milky Way, has a massive black hole at its centre; we are now looking at super-massive black holes, with masses thousands of times greater.

A black hole is a concentration of mass whose gravitational field is so intense that nothing can be emitted from it, neither any mass nor any radiation. Anything within a boundary, whose radius is defined by the total mass, cannot escape, while anything outside may be drawn in. Whatever is sucked in, whether it is part of a star, or a whole star, will disintegrate and become very hot before it disappears. The victim may be a star close to a black hole, the two making a binary pair in orbit round each other; the strong gravitational field strips off the outer layers of the normal star, which fall into the black hole. The black hole may be fed continuously from a surrounding disc of accreting material, or possibly whole stars may be sucked in, disintegrate, and disappear. Large amounts of energy are released in either case. The black hole itself cannot radiate, but the accreting matter can become very hot and radiate light and X-rays before it is swallowed up.

Black holes could in principle have any mass, but the varieties encountered in astronomy seem to be restricted to two main categories. At the centre of galaxies, possibly most galaxies, there are black holes with masses of several million times the mass of the Sun. The best known is at the centre of our Milky Way Galaxy; this has a well-determined mass of 4 million solar masses. On a smaller scale there are other, smaller black holes within our galaxy, which were discovered as powerful sources of X-rays. These all have companion stars in close orbit round them, and the energy feeding the X-ray emission is from material sucked in to the black hole from its companion. The distances and orbits of some of these X-ray binary systems are well enough determined to give accurate masses. The best known is Cygnus X-1, one of the first to be discovered; this black hole has a mass of 14.8 solar masses. All these binary black hole systems seem to be similar, with masses of between 10 and 20 solar masses. The galactic nucleus black holes are typically more than a million times more massive, and so far none has been found between these two categories.

Despite its enormous mass, or rather because of it, a black hole is very simply described. Whatever material it may have swallowed, no information about its contents can reach the outside world. Its only properties are its mass, its rate of rotation, and its electrical charge. The mass can be found from the dynamics of orbiting companions or, for the black hole in the centre of our galaxy, from orbiting stars (see Chapter 4). The spin is not so easily found, but it does affect the way in which matter falls into the black hole, which we describe later.

X-ray observations of the dynamics of this process suggest that the black holes in X-ray binaries are rotating very rapidly; the black hole in Cygnus X-1 is reported to be rotating 800 times per second.

Jets and Radio Lobes

In the 1960s, at the same time as the discovery of quasars, achieved by the Jodrell Bank interferometers and the optical spectra, Martin Ryle and his colleagues at Cambridge were developing new ways of building radio telescopes. New ideas combined with the rapidly developing power of digital computing led to telescopes with dramatically improved sensitivity and resolving power. Their new technique which became known as aperture synthesis transformed radio astronomy (see Chapter 10). Telescopes with multiple elements, building up apertures several kilometres across, could now be constructed. The first of these is now known as the Ryle 5 kilometre Telescope, which gave a resolution of 5 arcseconds. The map of Cygnus A made with this telescope is shown in Figure 31. It shows the structure of the twin lobes, and the central source. The two clouds are the twin lobes discovered in the first interferometric measurements of diameter, now seen to be complex structures fed from a central source located in the visible galaxy.

Aperture synthesis was evidently the way to make detailed maps with high resolution. Using wavelengths as short as 1 centimetre it became possible to map radio galaxies with a resolution of around 1 arcsecond, as good as any optical photograph available at that time. The outstanding radio telescope known as the VLA (the Very Large Array) was built in New Mexico during the 1970s, and completed in 1980. Plate 7 shows Cygnus A as mapped by the VLA at a wavelength of 6 centimetres. The small bright spot is at the centre of the galaxy which was discovered in the optical photographs. A jet leading from this central object to one of the lobes can be seen in the enlarged contour map (Figure 32). The whole

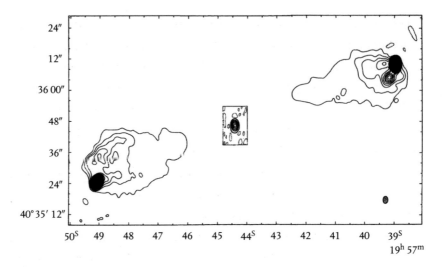

FIGURE 31 The Cygnus A radio galaxy mapped using the Ryle 5 km Radio Telescope at Cambridge in 1974. *Courtesy of Oxford University Press and the Monthly Notices of the Royal Astronomical Society.*

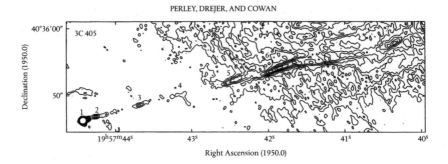

FIGURE 32 The jet in the Cygnus A radio galaxy. The central black hole is the origin of a jet whose energy is dissipated in the diffuse outer lobe of hot gas. VLA map at 6 cm wavelength. (Perley *et al.* 1984). *Reproduced by permission of the AAS.*

structure extends to over 300,000 light-years, which is more than 10 times the overall dimensions of our Milky Way galaxy.

Cygnus A is one of the most powerful radio galaxies known. Fortunately it is close enough (at a redshift distance of $z = 0.056$) to be seen in great detail; in contrast, most similar radio galaxies are much further away (in the region of $z = 1$). The VLA map shows that the twin lobes originate in an astonishing pair of jets, which stretch from very near the central bright spot all the way to the outer edge of each lobe. All the energy radiated by Cygnus A, at the rate of more than 10^{38} watts, originates in the central engine, and is fed through the jets to the outer edge of the clouds. This is more than the total energy radiated from a normal galaxy, and can only come from the gravitational pull of a very massive object. Furthermore, this central engine is so small that it can only be a black hole.

We do not understand exactly how gravitational energy is released as matter falls into the black hole, and how it is converted into an outward flow along the jets. Energy is dissipated in the disc of accreting material, which orbits the black hole at rapidly increasing velocity as it approaches the boundary of the black hole; in the last stable orbit the velocity can reach half the velocity of light. The total energy released can be a large fraction of the rest mass of the in-falling matter.

A key factor in the process is any rotation of the black hole, which can affect the accretion process. Without rotation, up to 6% of the rest mass of the in-falling material can be converted into energy, but with rapid rotation this fraction increases to 42%. This is the most efficient conversion of mass into energy anywhere in the Universe; in contrast, the conversion of mass into energy in nuclear fusion, as in the Sun, only achieves an efficiency of 1%. The rotation of the black hole also defines the direction of the jets, which emerge in two opposite directions along the spin axis of the black hole. Rotation also winds up the magnetic field into a straitjacket, which confines the jets as they leave the central engine.

Each jet contains very energetic electrons, positrons, and protons, streaming out at a velocity which may be as much as half the velocity of light. It blasts its way out from the galaxy containing the black hole, and out into intergalactic space, continuing on its way to become one of the largest structures in the universe. It remains intact until it encounters sufficient material in the thinly populated space between galaxies to halt its progress, where the streaming energy is converted into hot gas, forming a hotspot. At this point there is a shock front; the intergalactic matter is heated and streams out sideways, and particles can be accelerated to relativistic energies in the associated shock waves. It is the twin clouds formed around the hotspots from the jets which can be seen as the twin lobes of the radio galaxies. There is also a large magnetic field in and around the hotspots, with the result that the electrons emit the synchrotron radiation which is observed as the twin lobes. The clouds diffuse away from the hotspots, losing energy, and disperse after expanding sideways and, in some cases, diffusing about half-way back towards the centre.

The twin jets are nearly, but not exactly, in line with each other. It seems their directions may change by a small amount on a time scale of about a million years, an idea that is supported by the appearance of multiple hotspots. The theory is that a jet acts like a dentist's drill, punching through the cloud until it is stopped at the shock front. When the jet moves sideways, the old hotspot decays slowly, and remains visible alongside the new one.

The same pattern of a central engine, twin jets, and diffuse lobes is seen in maps of many other radio galaxies. Plate 8 shows two examples in which there are twin jets but no hotspots. In NGC 1265 the jets are bent, swept back by the ram pressure associated with the high velocity of the parent galaxy through the intergalactic gas of a cluster of galaxies. There are also many radio galaxies which appear to have only one jet; two examples are shown in Figure 33 and Plate 9.

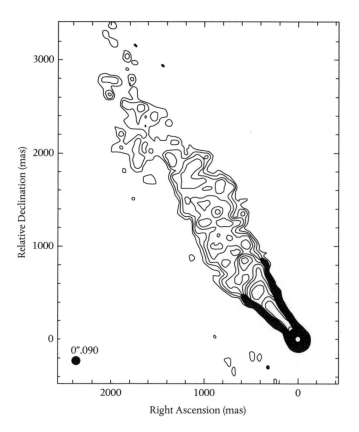

FIGURE 33 A radio galaxy with only one jet. 3C 264 MERLIN + EVN. There may in fact be a second jet, which does not appear due to a relativistic beaming effect. *Lara, L. et al., Astronomy & Astrophysics, vol 415, page 905, 2004, reproduced with permission © ESO.*

All One Family

The quasars and the radio galaxies at first appeared to be different classes of object, with the quasars distinguished by their single bright star-like centre and the radio galaxies by their jets and radio lobes. But they both have similar very high energies, and both are seen at similar very great distances. They are in fact two manifestations of the same basic phenomenon, with a black hole as the central engine.

The difference is only in the radio brightness of the central engine, which in the radio galaxies is hidden from view. The link between them was first established by the observation, already mentioned, that 3C273, one of the brightest quasars, has a prominent jet, very much like the jets of the radio galaxies depicted above.

In 3C273 there is apparently no jet on the opposite side. Could this be the same phenomenon as the one-sided jets in many radio galaxies? The theory is that there are in fact two jets, one which is visible pointing almost directly towards us, and the other invisible pointing away from us. Both jets would produce a cloud of emission, like the twin clouds of Cygnus A; seen end on, these clouds would appear superposed on the jet. Indeed there is such a faint extended cloud superposed on the jet. But why is the direction of one jet making it visible, and the direction of the other making it invisible? The solution involves a remarkable example of the Special Theory of Relativity.

Faster than Light?

The jets of several quasars contain a series of concentrations, as though the central engine has shot out separate packets of energetic electrons along the same channel. Over a time scale of only a few years, a repeated set of maps using the high angular resolution of VLBI (Very Long Baseline Interferometry) shows these blobs are moving along the jet. Figure 34 shows two examples.[34] We know how far away each quasar is, so we can calculate the velocity of the blobs. The astonishing result is that they seem to be moving faster than the speed of light, c, in fact as much as double the speed of light. This 'super-luminal velocity' is an illusion, which can be explained by a simple model due to Martin Rees.[35] If the actual velocity is somewhat less than c, but the track along the jet is directed almost towards us, the velocity as seen by us can appear to be many times greater than c. This due to the compression of time for an object moving towards a stationary observer.

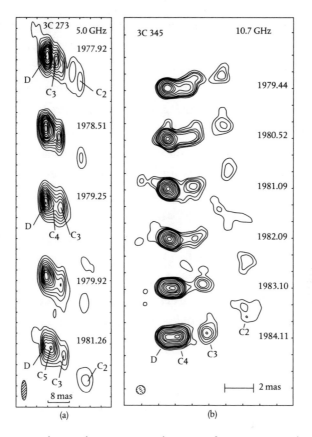

FIGURE 34 Superluminal motion in the jets of 3C273 (a) and 3C345 (b). *(b) Reproduced by permission of the AAS.*

Velocities approaching *c*, which are called 'relativistic' velocities, also have an important effect on the strength of the radiation from the electrons streaming along the jet; their radiation is concentrated towards the direction of travel, forming a kind of lighthouse beam pointing along the jet. This effect is known as relativistic beaming; we will encounter it again in the story of pulsars (Chapter 6).

Relativistic beaming accounts for the apparent one-sided jet in many quasars, such as 3C273. Even when both jets can be seen, as in radio galaxies such as Cygnus A, there may be a degree of relativistic

101

beaming. Although the radio maps such as that for Cygnus A appear
to show jets oriented perpendicular to the line of sight, they could be
tilted at an angle up to 45°. There would then be a brightness ratio
between the two jets due to mild relativistic beaming, the brighter jet
having a component of velocity towards us. So, if the quasars and
radio galaxies are fundamentally the same, differing only in their
alignment, how is it that the radio galaxies have little or nothing of the
bright core which dominates the quasars? Figure 35 shows the geometrical model which explains this.

In this model the black hole, which is the source of the gravitational
field sucking in stars and gas, is surrounded by an opaque doughnut-
shaped cloud. This torus is formed by the accreting material, which
does not fall in directly but circulates around the black hole in collapsing orbits. The axis of this torus is the same as the direction of the jets,
and is the same as the axis of rotation of the black hole. The hole in
the torus allows us to see inside if our line of sight is within an angle

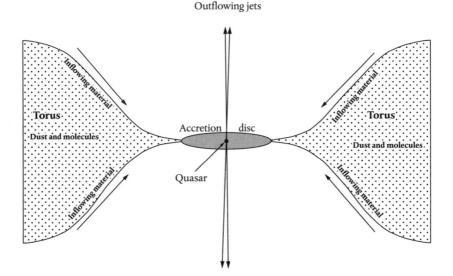

FIGURE 35 Unified model of the cores of radio galaxies and quasars.

of around 45° of the jet. Furthermore, if we are looking closely along the direction of the jet, relativistic beaming will emphasize the jet on our side, and weaken that on the other. Here is a unifying model which allows us to say that radio galaxies and quasars are simply different views of the same phenomenon.

Many observational tests have been made of this model. One radio test is to look at the small radio signal from the central engine which remains as it penetrates the torus, accounting for the bright spot in the centre of Cygnus A. The torus contains neutral hydrogen which, as we have seen in our Galaxy (Chapter 3), has a spectral line at a radio wavelength of 21 centimetres. Radio from the central engine is absorbed at this wavelength in the torus, giving a measure of the amount of hydrogen and at the same time a measure of the spread of velocities in the line of sight. The results fit the model remarkably well.

Gravitational Lenses

In an early survey at Jodrell Bank Observatory of quasars, intended to provide sufficiently accurate positions for them to be identified, Dennis Walsh noticed a pair of quasars of almost the same radio brightness, only 5.7 arcminutes apart.[36] This might have been a chance coincidence, but when it was found that the optical spectra of both had the same large redshift ($z = 1.4$), it was realized that this was the first example of the phenomenon of gravitational lensing. The two quasars were in fact one and the same, but they were seen along two different ray paths.

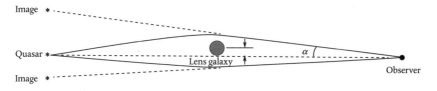

FIGURE 36 Ray paths in a gravitational lens.

Figure 36 shows the situation. The massive object labelled M, which is a whole galaxy, lies between us and the quasar. According to the theory of General Relativity (see Chapter 8), the gravitational field of this object distorts space around itself, so that rays of any type of radiation which travel in straight lines now follow lines in curved space. All rays from the object O behind M bend towards it, in a similar action to an optical lens, but with a different geometry. the bend in these lines is greatest for rays nearest to the massive object (which is

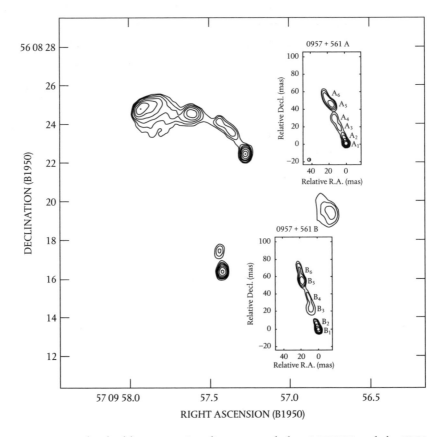

FIGURE 37 The double quasar. A radio map made by eMERLIN and the EVN showing details of the two main lensed components. *Courtesy Tom Muxlow and University of Manchester.*

the opposite of rays in a conventional lens). In the figure there are two rays, both bent by the lens but on opposite sides, accounting for the double quasar. This strange geometry can produce multiple images; in theory there should be an odd number, one, three, or five, although they would have different brightness and some might be too faint to see. A near alignment of M and O can produce a magnified and brighter image, a situation which has been exploited in mapping some very distant quasars. Figure 37 shows the extended image of the double quasar mapped in detail by the MERLIN interferometer network. Here the two bright images, shown enlarged, each display the complex structure of the distant quasar, with a bright central source and a jet. The lensing object is not an ideal point mass, but a whole galaxy, whose complex structure adds complication to the images.

Gravitational lensing requires a chance alignment of quasar and massive galaxy which can happen only occasionally. After its discovery in radio, it was found in optical photographs of distant quasars and galaxies. The lensing object is often a complete cluster of galaxies, with the effect that many objects are distorted into magnified and extended images. This effect is useful in exploring the population of very distant and faint objects, but also in finding the total amount of gravitating mass in the intervening cluster.

6

Supernovae and Pulsars

Of the many astronomers who have opened the gates of the universe outside our Galaxy, Fritz Zwicky (1898–1974) must surely be the outstanding leader. He was a tireless and expert observer of supernovae in extragalactic nebulae; he was the first to understand this phenomenon, as he correctly claimed in an unconventional preface to his *Catalogue of Selected Compact Galaxies*, published in 1968:

> In the Los Angeles Times of January 19, 1934, there appeared an insert in one of the comic strips, entitled 'Be scientific with Ol' Doc Dabble' quoting me as having stated 'Cosmic rays are caused by exploding stars which burn with a fire equal to 100 million suns and then shrivel from ½ million miles diameter to little spheres 14 miles thick' says Prof. Fritz Zwicky, Swiss Physicist.' This, in all modesty, I claim to be one of the most concise triple predictions ever made in science. More than 30 years were to pass before this statement was proved to be true in every respect.

There was of course a more conventional publication of these ideas later in 1934, by Zwicky and Walter Baade (1893–1960) in the *Proceedings of the USA National Academy of Sciences*, but Zwicky was correct in his claim. The neutron had been discovered by James Chadwick

(1891–1974) only 2 years previously, and here was a prediction that a whole star with the mass of the Sun could be made only of neutrons, and packed into a sphere only 22 kilometres in diameter. This neutron star would be the result of the collapse of a normal star under its own gravity, when its nuclear energy supply was used up. The collapse would be spectacular, and visible as a 'supernova'. Furthermore, the event would release so much energy that it could generate the cosmic rays that pervade our Galaxy, whose origin was still unknown. As Zwicky remarked, all these predictions have been proved correct, even in the diameter of the neutron star (although it later took some sophisticated theory to establish this unequivocally).

Radio astronomy comes into this story in two ways. The cosmic rays which figured in the original prediction turned out to be the origin of the radio emission from interstellar space, discovered by Jansky and understood eventually as synchrotron radiation (Chapter 3). It is still difficult to associate all cosmic rays with supernovae, but the evidence is getting stronger, as found recently by the Planck satellite (Chapter 9). The main link with radio astronomy, however, was the neutron star itself. This was thought to be virtually unobservable, although it might still have enough heat remaining from its energetic collapse to be a faint but detectable X-ray source. Theorists agreed it should exist, but no-one predicted that it would advertise its presence with a flashing light, or as a radio beacon that became known as a pulsar.

Jocelyn Bell's Discovery

It is fair to say that most early radio astronomers started with a background of communication engineering rather than nuclear physics, and the idea of a neutron star was not in their minds when the first pulsar was discovered. At Cambridge, UK, Antony Hewish and his student Jocelyn Bell (b. 1943) built a large antenna array to observe small bright radio sources, and particularly quasars. They were interested in the way the radio signal varied, sometimes rapidly during an observation lasting only a few minutes. These fluctuations in intensity of radio sources were known to occur as the radio waves passed through our Solar System. Interplanetary space is not quite empty, but contains ionized gas flowing out from the Sun. Irregularities in this plasma affect radio waves by refraction and diffraction; the effect is known as *interplanetary scintillation*. It is usually observed at low radio frequencies, and it is closely related to the effect in the Earth's ionosphere described in Chapter 3. Hewish and Bell had two objectives. First they intended to monitor the outflow of ionized gas from the Sun, by observing scintillation in known quasars at various distances from the Sun, and how it varied with solar activity. They also knew that a distant radio source would only scintillate if it was small, like a quasar, and not spread out like a radio galaxy. Observing whether a source was steady or fluctuating would sort out the quasars from the radio galaxies. The antenna had to be large enough to record rapid fluctuations in a number of sources spread over the sky. It was constructed as an array of 2048 dipoles at 3.7 metres wavelength, covering an area of 1.8 hectares (4.5 acres). Like many research students at Cambridge, Jocelyn Bell constructed her own apparatus, including the antenna, and spent hundreds of hours analysing the voluminous records produced by the receiver.

The unique approach of this experiment was to record fluctuations as rapid as one-tenth of a second, in contrast to the usual practice in radio telescopes of integrating and smoothing the signal for many seconds or minutes to obtain sufficient sensitivity. The signals appeared as pen recordings on several metres of paper each day, showing the transit of a series of radio sources, with various degrees of fluctuations due to scintillation. Soon after the start of observations in July 1967, Bell discovered a strange source that seemed to be all fluctuations and no steady signal; furthermore, it was very variable in strength, and did not appear every day. A repeat observation of this radio source in November, using a receiver with an even shorter response time, revealed the first pulsar signal (Figure 38), in the form of a series of short pulses at a precise interval of 1.334 seconds. The discovery won Hewish a Nobel Prize in 1974. This pulsed radio source, now known as PSR B1919+21,[37] was a total surprise. At first it was thought it must be man-made, either terrestrial interference or a transmission from some orbiting satellite; alternatively it might even be the first detection of extra-terrestrial intelligence. The latter thought was disturbing, and conjured up visions of the world's press clamouring for the first news of the 'little green men' who were trying to communicate with us. The observers decided to keep quiet until they had worked out what was happening. Two months later they were ready.

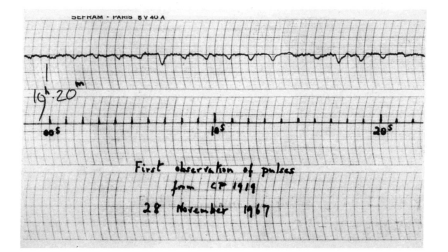

FIGURE 38 The first pulsar recording. The pulses appear at exact intervals of 1.337 seconds.

The *Nature* letter of February 1968 which announced the discovery contained a remarkably penetrating analysis. The very short periodicity seemed to rule out any source in a large object such as a star, but at the same time the astonishing precision of the periodicity soon allowed a crucial test which showed the source was outside the Solar System. Watching the arrival time of the pulses as the Earth moved round the Sun, even for only a month or two, could detect whether the source was in orbit round the Sun, since there would be a varying delay in pulse arrival time with the changing distance from Earth. The only variation actually observed was a very obvious effect of the Earth's orbit itself, which gave a large annual variation of several minutes, comparable with the 8 minutes' travel time for light to reach the Earth from the Sun. The amplitude and phasing of this annual variation gave an accurate position for the source even after only a few months. By February, Bell and Hewish had found three more pulsating sources, one with the even shorter period of a quarter of a second, and they were well on the way to finding accurate positions for those also. There was nothing to be seen on any sky photographs at any of the positions. What could these strange objects be?

The *Nature* letter mentioned the right answer, a neutron star, but only among several possibilities. Several types of star are known to oscillate, but only much more slowly, with periods of hours or days. A very small star, such as a white dwarf, might be ringing like a bell, which would possibly give a precise period of around 1 second. The unfamiliar neutron star, known only in theory, was the only object small enough to fit the very short time scale. Theorists immediately got to work, calculating modes of oscillation that could account for the pulsation. They were on the wrong track.

The Crab Pulsar

The solution to the problem had, in fact, already been published by Franco Pacini (1939–2012) a few months earlier, in a letter to *Nature*. It was well known that the Crab Nebula, a remnant of a supernova explosion, contained a source of energy which maintained its emission of light and X-rays. Pacini suggested that this might be a neutron star at the centre of the nebula, the remnant of the stellar collapse as suggested by Zwicky. If the neutron star was strongly magnetized, and if it was rotating rapidly, it would act like a generator in a huge power station. The whole energy radiated by the nebula would come from the kinetic energy stored up in the rapid rotation of the star. Furthermore, he predicted that the star itself might be detectable as a radio source.

Pacini's suggestion was followed by another *Nature* letter, from Thomas Gold (1920–2004), that set out very clearly the case for identifying pulsars with rotating neutron stars (Gold had evidently not known of Pacini's work, even though at the time Pacini was on a visit to Cornell University, and was sitting in an office almost next door to him!). The key contribution in both papers was the simple test: if the energy came from rotation, then the pulsar would be seen to slow down, while if the pulsation was due to an oscillation, it would not. The slow-down would be most obvious in a rapidly rotating, young pulsar, and as it happened the ideal candidate turned up shortly afterwards in the Crab Nebula itself.

Radio pulses from the Crab Pulsar were discovered in the USA by David Staelin (1938–2011) and Edward Reifenstein in 1968. Although they were extremely variable in amplitude, they soon established a periodicity of 33 milliseconds, and the slow-down was found soon afterwards. The power station had been identified. As Pacini had pointed out, it was a spinning generator, like the electricity generator in a terrestrial power station except that it was spinning once every 33.2 milliseconds, producing power at 30.2 Hz rather than 60 Hz (the main electricity supply in the USA), and with a rotating magnet over a million times stronger than any magnetic field attainable on Earth.

The electric field generated by the rotation of such an enormous magnetic field is large enough to extract electrons from the surface of the star, creating an ionized and magnetic atmosphere in which the radio waves are generated. A classic analysis in 1969 by Peter Goldreich (b. 1939) and William Julian (b. 1939) of this *magnetosphere* gave a reference model which still stands today. The magnetosphere is attached to the rotating star by the strong magnetic field; it co-rotates and extends outward to such a distance that its speed is close to the speed of light. Conditions inside the magnetosphere are extreme and unknown anywhere else.

Hunting for Pulsars

As soon as they heard of the discovery, radio astronomers all round the world joined in with whatever resources were available. In Australia the east–west arm of the Mills Cross, a mile-long cylindrical reflector, was used to survey the southern sky; among others, a pulsar with the short period 89 milliseconds was found in the young supernova remnant in the constellation of Vela. At Jodrell Bank, we had the advantage of the fully steerable 78-metre telescope now known as the Lovell telescope; this turned out to be the ideal instrument for following pulsars across the sky and investigating details of their pulses. Papers flooded in to *Nature*, the first arriving from Jodrell Bank only 3 weeks after the discovery paper. The wavelength range was soon found to extend from the long wavelength of the discovery, 3.7 metres, to as short as 11 centimetres, as observed at Owens Valley Radio Observatory in the USA. The precise nature of this broadband emission was unclear (and remains so to this day), but the hunt was on for more pulsars and any associations that could be found with known visible stars or nebulae.

Surveying the sky in a search for pulsars turned out to be a difficult technical challenge. The signals are weak, and they must be detected against an inevitable background of electrical noise, part of which is generated inside any radio telescope and its receiver system, and part of which is the cosmic radio noise originally detected by Jansky. Distinguishing a genuine signal from the background noise is a problem familiar to radio astronomers. The signal itself must be made as large as possible; this means that only a large radio telescope can be useful in the search. Next the broadband nature of the signal can be exploited by using a broadband receiver to collect as much as possible of the wanted signal; in modern searches a very wide bandwidth of 500 MHz or more may be used, giving an advantage over the bandwidths of several megaherz that were available in early searches. The problem then is *dispersion*, a phenomenon in which the pulse arrival time varies with frequency, giving a wide spread over such a wide frequency band.[38] Unless dispersion can be cancelled out, the pulses may be smoothed out and lost. The next expedient usually adopted in sensitive radio observations is to integrate the signal over a long observing period of many seconds, minutes, or sometimes even hours; this again was inappropriate as it would smooth out the pulsed signal by which the pulsars could be recognized. Both these problems can be overcome if the pulse period and the dispersion are precisely known. Catalogued pulsars are routinely observed using sophisticated signal processing, but a search for new pulsars with unknown characteristics is very demanding.

The Big Search

The majority of the known pulsars, which now number over 2000, have been discovered in a remarkable collaboration between UK and Australian radio astronomers. In 1976, Andrew Lyne from Jodrell Bank Observatory, spent a year at CSIRO Radiophysics in Sydney working with Richard Manchester. Their first survey[39] used the same east–west arm of the Mills Cross that had already found the Vela pulsar and several others. This long cylindrical reflector has a very large collecting area, which was ideal for a blind search for pulsars with unknown positions and periods. It needed improvements, and Lyne built and installed new low-noise amplifiers, while Manchester organized a new scanning method. Each new detected, or suspected, pulsar was then followed up with observations on the fully steerable 210-foot (64-metre) radio telescope 300 kilometres away at Parkes, which was equipped with a new generation of receivers and computers. The collaboration extended to Joseph Taylor of the University of Massachusetts, who was an expert on signal processing (Taylor later won a Nobel Prize for the discovery of the Relativistic Binary Pulsar; see Chapter 7). The combination of the expertise of these three, and the combination of the two different radio telescopes, led to the discovery of 150 new pulsars. Manchester and Lyne then moved on to an entirely new approach, using only the Parkes telescope. Their new approach depended on the rapid developments in digital computers, and on a development of the Parkes telescope which enabled it to work 13 times faster.

A conventional reflector telescope, with a single antenna feed at its focus, has a single narrow beam in which it can pick up a radio signal. The larger the telescope the more sensitive it is, but also the narrower is the beam. A large telescope is necessary in a pulsar survey, because the signals are weak, so scanning the sky for weak pulsars has to use a small beam and is a very slow business. The solution at Parkes was to install a multiple-feed system, making a cluster of 13 beams. Each feed had two low-noise amplifiers (for two polarisations), cooled to 20° above absolute zero, and covering the very wide bandwidth of 300 MHz centred on 1.4 GHz, a wavelength of 21 centimetres. It was this Parkes Multibeam Pulsar Survey that brought the total of known pulsars up to close to 2000.

There remained the problems of dispersion and integration. To overcome the effects of dispersion only a narrow bandwidth could be used, but sensitivity demanded a wide bandwidth. The solution was to divide the wide band into many separate narrow receiver channels, each with its own detector system; this had to be done for each of the 13 beams, making a total of 1248 individual receivers. The sensitivity of the survey depended on integrating the radio signal; each patch of sky was observed for half an hour or more; during this time the signal in each channel was recorded at intervals short enough to follow the pulses from pulsars which might have periods as short as a few milliseconds. The result was a mountain of recorded data, apparently consisting only of random noise but with signals from unknown pulsars embedded within it. This became easier to handle as new digital recording media became available, and the subsequent analysis also became more efficient with the development of new computer systems. A wide range of both dispersion and periodicity had to be searched in each beam at each telescope pointing. There were 3000 telescope pointings, and each of the 13 beams for each pointing needed 2 hours of computer time. It would have taken a single computer working full time for 9 years to complete the analysis; in fact the task was shared among several computers in several observatories in several countries.

The original Parkes survey has been followed by several others using the same basic system, but much faster computers. All the raw data still exist as recorded digitally, and new computer methods are still being devised to extract from the recordings some originally overlooked pulsars. Other telescopes, such as the Green Bank 100-metre GBT telescope, have been used to search particularly interesting regions of the sky, such as the region of the Galactic Centre, but it seems unlikely that another survey covering most of the sky will be carried out until new telescopes like LOFAR and the Square Kilometre Array come into operation (Chapter 11).

Hunting Techniques

Only a few pulsars give strong enough signals for their pulses to be seen on simple recording systems like the array used by Hewish and Bell. Building the catalogue of around 2000 pulsars has required some increasingly sophisticated methods for sorting out periodic pulses from an inevitable background of noise, generated by the galactic background and in the local receiver system. The classic method of detecting a periodic signal against a noise background is spectral analysis, using Fourier techniques familiar to mathematicians, physicists, and engineers. Suppose a record of receiver output, typically lasting many minutes, and which looks and sounds like noise, contains a low-level periodic signal in the form of a pure sine wave. The analysis technique is to multiply the recorded output by a locally generated sine wave of the same frequency, which will pick out the required signal. Multiplying by sine waves with any other frequency will only emphasize part of the noise, so a plot of the product against the frequency of the local sine wave will display the spectrum of the input signal. The result is a spectrum which will show a peak at the pulsar frequency. This is evidently a tedious repetitive process which requires a digital computer, preferably one with a high speed and the capacity for handling many receiver channels simultaneously.

A closely related method is known as autocorrelation. Here the output signal is multiplied by a copy of itself, the copy being delayed in time by a variable amount. If the delay is exactly equal to the period of the low-level signal, the signal is picked out in the product, and the spectrum can be plotted by repeating the process with a variable amount of delay. Since a full-scale search may mean looking for a pulsar with any period from a millisecond to 10 seconds, the process requires repetition over a full set of delays, and it requires careful organization to avoid overwhelming the available computer capacity. Both the Fourier analysis and the autocorrelation methods can be refined to search for narrow pulses rather than sinusoids, and both demand high-speed digital computations to keep up with the many channels of noisy data that a typical search must produce.

Gamma-Rays and Photons

Radio waves are at one end of a very wide spectrum of electromagnetic waves emitted by pulsars. Although a large proportion of the energy in their pulses is emitted at the high-energy end of the spectrum, in X-rays and gamma-rays, such radiation does not penetrate the terrestrial atmosphere, and it must be detected using receiving systems in spacecraft. At the high energies of X-rays and gamma-rays, radiation from any cosmic source arrives on Earth as discrete photons, each containing a large amount of energy and easily detected but only arriving very occasionally. The situation is especially important at the highest energy gamma-rays; even using the large orbiting gamma-ray telescope which is mounted in the Fermi satellite observatory, there may typically be only a few thousand photons detected from a pulsar in a whole year. Although these photons appear at first to be arriving at random time intervals, they do in fact only arrive within the same time pattern as the radio pulses. Gamma-rays from a pulsar with a known period can therefore be detected by the same signal processing methods, despite the very long observation time needed to collect enough photons. Discovering new pulsars, with unknown periods, is almost impossible using the same techniques as in radio. But new pulsars have been discovered in gamma-ray pulse arrival times recorded over a year or more. This analysis was made by using the intervals between pulse arrival times. These intervals must be at discrete multiples of the pulse period, which can be found by a similar Fourier or autocorrelation search technique. Several previously unknown pulsars have been discovered in this way.

How Do Pulsars Work?

Part of the excitement of working in the field of pulsar research is that most of the many discoveries were unexpected or not even contemplated, apart from the original prediction by Zwicky of the existence of neutron stars. No-one suggested that neutron stars would advertise their existence with a radio transmitter, or that such a transmitter might be beamed or that it would be swept round the sky like a lighthouse at the rate of 1 rotation per second or even faster. In fact, even now no-one completely understands how this happens. The idea of a neutron star, and even its size and mass had been predicted remarkably accurately, but the possibility of observing such an insignificant object remained remote.

When pulsars were discovered, the internal structure of a neutron star was already a matter of some interest to physicists working in the field of condensed matter. Conditions inside a neutron star were beyond the limit of experience and theory; in a neutron star particles are closer together than their closest approach when colliding together in the highest-energy accelerators which can be built on Earth. What controlled the forces between neutrons packed so tightly together? There were useful estimates of the mass and size of such objects; they should be about the same mass as the Sun, but only about 20 kilometres across, about the size of a city. But they were still thought of as a simple mass of neutrons, like one large molecule.

The observations soon pointed to concepts which were new to astrophysics. Most of the star must be a fluid rather than a solid, and only an outer crust is a crystalline solid. The fluid is in an extraordinary state only previously known in small-scale laboratory experiments in low-temperature physics; it is a 'superfluid' and a 'superconductor'. A superfluid can flow without any viscosity, and a superconductor can carry an electric current without any resistance. Both effects occur only at high density and low temperature, when the interaction between fundamental particles overcomes their random movements. In the neutron star the density is so high that the temperature hardly matters; it could be as high as 10^6 K and the interior will still be superfluid. Classical physics no longer applies; the behaviour can only be described by the laws of quantum mechanics.

The crust is a super-strong crystal, made of nuclei of iron and other massive elements, containing free electrons and neutrons. The whole structure sustains the strongest magnetic field known anywhere in the Universe, generated by electric currents flowing without any resistance. The dynamics of the rapid rotation of such a system had not been contemplated, although it was known that rotating a superfluid involved some very odd physics. What actually happens was only revealed by extended and patient observations of the rotation by observing radio pulse arrival times, and by measurements of the detailed characteristics of the radio pulses.

As we have seen in the previous chapter, an important prediction by Gold and by Pacini was made soon after the discovery of pulsars; who both pointed out that if the energy feeding the radiation from the Crab Nebula came from the rotation of the pulsar at its centre, that rotation would be observed to slow down on a time scale of the order of a thousand years, the age of the Nebula since its origin in a supernova explosion. The slow-down was easily observed both in the Crab pulsar and in the Vela pulsar, another young pulsar discovered by Australian radio astronomers Michael Large and Alan Vaughan in the first survey of the southern sky. The slow-down rate depends only on the strength of the magnetic field and on the rotational inertia of the star, which was already estimated to be, by coincidence, approximately the same as that of the Earth (the star has a much greater mass but a much smaller radius). The magnetic field was shown to be around a trillion (a million million) times that of the Earth.

The Glitch

A few weeks after the first observation, the rotation of the Vela pulsar interrupted its steady slow-down with a sudden speed-up. Timing the rotation of most other pulsars showed that this is a common phenomenon, which became known as a 'glitch'. The effect is easily seen, even though it is small; the step in rotation rate in various pulsars is between one part in a million and one part in a billion.

The physics of the glitch illustrates the extraordinary conditions in the interior of these neutron stars. The two components, the solid crust and the liquid core, can rotate independently, and only become locked together at the time of the glitch. Between glitches the crust slows down normally, but because the core is superfluid it stays spinning independently, at a constant rate. There is almost no friction between these two components. As the crust slows down the difference in rotation speed increases to a point where the two components suddenly lock together, slowing down the core and speeding up the crust. The independent rotation and the locking mechanism could only occur at a low temperature when quantum physics prevails. Physicists say that an object which is so cold that it has lost its usual properties has become 'degenerate'.[40] So what is it about rotation in such a degenerate object that allows the two components to behave independently and then suddenly to lock together?

If a pot of superfluid helium is rotated in the laboratory, the liquid does not rotate as a whole but separates into discrete vortices; the faster the rotation, the more vortices appear. The density of vortices is proportional to the speed of rotation, so that as the rotation slows down, vortices must migrate outwards, so reducing the density. The whole of the neutron star core behaves like this, although only part of the superfluid is involved in the glitch. This part co-exists with the inner part of the crust, and here the vortices are normally attached firmly to the crystal lattice of the crust. The vortices cannot move, so that the density of vortices in this component is fixed, which means that the speed of rotation is fixed. Perversely, the pattern of vortices in the superfluid is locked to the crystal structure in the crust, but it cannot slow down along with the slow-down of the crust. The two rotation speeds, of the crust and the superfluid, build up a difference, which eventually becomes so large that the vortices become unpinned; they then suddenly move outward to reduce their number density, and the rotation difference disappears. The difference in rotation rate then builds up again, over a period of several years, until another glitch occurs.

In young pulsars these glitches occur at irregular intervals of a few years, while in the older 'recycled' pulsars, which we deal with in the next chapter, the interval may be thousands or even a million years. The statistics of this behaviour do not exactly fit any simple pattern, although there is a good analogy in the so-called sand-pile effect, in which a steady stream of sand builds up a pile which suddenly collapses when it reaches a critical height, and builds up again after a randomly variable time. Between glitches, most pulsars behave very smoothly; pulsars would be among the very best clocks in the universe if they did not suffer from these irregular glitches.

The Biggest Magnet in the Galaxy

Stars and planets all have magnetic fields. Earth has a magnetic field of about 1 gauss. Our Sun, which is a fairly typical star, has a simple dipole field of about 50 gauss extending through the whole volume, with extra local complex and variable components which occasionally surface in sunspots. The collapse of a star at a supernova condenses its extensive dipole field into a small area, increasing the field strength by 10 orders of magnitude. Newborn neutron stars have dipole fields measured in teragauss (10^{12} gauss). These huge fields are sustained by electric currents in the interior. Neutrons themselves cannot be responsible for the current, but about 10% of the neutron population is split into electrons and protons; the electrons carry the current which produces the magnetic field. Furthermore, the electrons are superconducting (nothing is normal inside a neutron star!), so that the current continues indefinitely without diminution. There is no obvious difference between the field strength of young and old neutron stars, even on the time scale of a million years.

As in the familiar bar magnet, the magnetic field lines around a neutron star form a pattern like that of Figure 39. The pattern is rotating rapidly, like the rotor of a dynamo, and enormous electric fields are generated outside the star. These electric fields are strong enough to extract electrons from the surface and accelerate them to very high energies. The space around the star becomes filled with charged particles, forming an extended magnetized atmosphere, the 'magnetosphere', which rotates with the solid star. Furthermore, the magnetosphere extends so far outwards that its outer parts must move very rapidly to keep up with the rotation. The limit is reached when this co-rotation speed reaches the speed of light, at a radial distance of some hundreds or thousands of times the radius of the star. Within this region are generated all the signals we receive from a pulsar, from radio through optical and X-rays to gamma-rays.

Generating Radio and Gamma-rays

Two regions of the magnetosphere radiate the beamed signals which tell us so much about the pulsars. The magnetic field dominates and determines the structure of the magnetosphere. Immediately outside the polar region, shown in Figure 40, there is a near vacuum in which electrons are accelerated to very high energies. Each electron interacts with the strong magnetic field, radiating a gamma-ray which generates a further pair of particles, an electron and a positron. These in

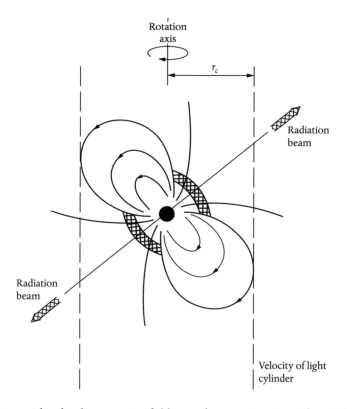

FIGURE 39 The dipole magnetic field round a neutron star. The pattern is confined within a 'velocity of light cylinder', within which the whole pattern rotates with the neutron star. A large electric field above the polar region accelerates charged particles to very high energies. Radio and other radiation originate above the polar region and in a vacuum gap at the boundary of the polar region. Within the equatorial regions the particles are confined to the field lines and no radiation is observable. The field lines from the polar regions penetrate the cylinder, and charged particles (electrons and positrons) can flow outwards.

turn form a cascade, stimulated originally by the electrons but now including positrons. Both are so energetic that they can radiate gamma rays. This region of the magnetosphere is called the polar cap. The polar cap joins at its edges on to the second high-energy region, the outer gap, which extends out to the limits of the magnetosphere. This

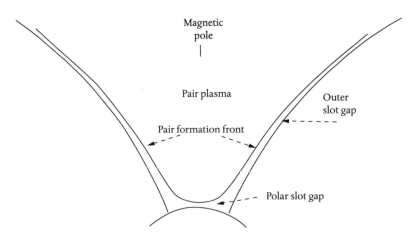

FIGURE 40 The polar cap, and the outer gap. The radio emissions from pulsars originate in these two regions of the neutron star magnetosphere.

outer vacuum gap, in which the highest electric field is maintained, is shown in Figure 40 as a narrow region at the boundary between the polar and equatorial regions of the magnetosphere.

Throughout the magnetosphere the magnetic field determines the outward path of the charged particles, and as they travel outwards they radiate in a narrow beam, which is the lighthouse beam observable in radio, light, and gamma-rays. The shape of the beam reveals the origin of the radiation, which for gamma-rays is mainly in the outer gap, while radio is generated somewhere over the polar cap. For some young pulsars, notably the Crab Pulsar, we can see radio generated in both regions.

The intensity of the radiation in optical and gamma-rays is directly related to the energy of the particles, as it is in the case of the radio emission from the solar corona and from the Galaxy. But most of the radio emission comes from a far more efficient process, which takes place in a small region near the base of the vacuum gap and extending over the whole of the polar region. Here the radio emission is due to the coherent motion of many electrons or positrons, as it is in solar flares or indeed in any terrestrial radio transmitter. How this works is not understood: it is embarrassing for astronomers to have to admit that neutron stars were only discovered because they radiate by a process that is still not understood after more than half a century.

The Lighthouse Beams

Every pulsar has its own individual beam shape, which indicates the origin of its lighthouse radiation. There is an extensive library of recorded pulse shapes, and many papers have been written on their interpretation. Some pulsars radiate detectable beams in radio, visible, X-rays, and gamma-rays, and the pulse shapes may be different in each band. The best example of this is the Crab Pulsar, whose radio, optical and gamma-ray pulse shapes are shown in Figure 41.

The double peak in these pulse profiles indicates that most of the radiation is generated in the outer gap of the magnetosphere, which is in our line of sight twice in each rotation of the pulsar. This radiation is satisfactorily explained as directly generated by charged particles with very high energy, streaming out through the outer gap. At a low radio frequency, as seen in Figure 42, there is an extra component, which has a different origin; it is generated above the polar cap. In most pulsars we can see only radio emission, which is from over the polar cap.

How Does Pulsar Radio Work?

Radio transmitters on Earth work by generating oscillating electric currents which flow through various types of antennas, from the tiny wire loops inside your mobile phone to the dipoles and horns which we see mounted in dish reflectors on tall masts. Pulsars certainly generate very large electric currents, which are powerful enough to radiate the pulses we see in light, X-rays, and gamma-rays. At these high-energy regimes the radiation is simply the sum of the radiation

ABDO ET AL.   Vol. 708

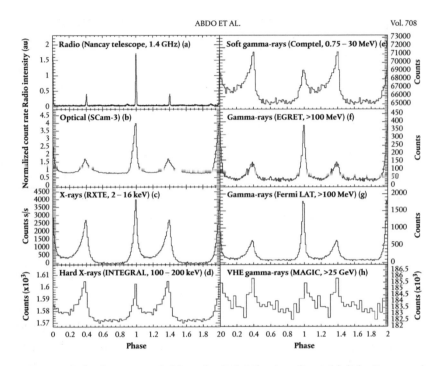

FIGURE 41 Pulse shapes recorded from the Crab Pulsar in radio, visible light, X-rays, and gamma-rays. All these profiles are shown for two complete periods. (Abdo *et al.* 2010). *Reproduced by permission of the AAS.*

from a multitude of individual electrons as they stream out along the outer vacuum gaps of the magnetosphere. Pulsar radio transmitters are entirely different. At this low energy, long wavelength part of the spectrum, the electrons in an efficient transmitter have to move together, as they do in a terrestrial transmitter. This means that there must be some sort of resonance, in which electrons move coherently. It follows that any particular radio frequency must be radiated in part of the magnetosphere which is in some way tuned to that frequency. The only means we have to track down the seat of these resonances is to look at the shape of the radio beams.

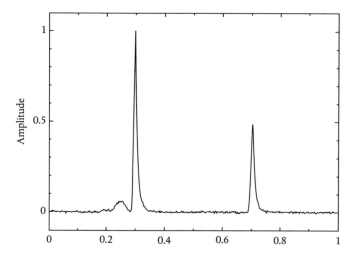

FIGURE 42 The Crab pulsar profile at a low radio frequency (610 MHz), showing a precursor pulse.

These shapes of the radio pulses are cross sections of the radio beams, which cross our line of sight once for each rotation. The pulse covers a fraction of the rotation period, showing that the beam is a few degrees wide. This width is an indication of the width of the transmitting region in the magnetosphere. The pulse is wider at lower frequencies. The idea is that radio travels out from the transmitter along the magnetic field lines, which splay out above the magnetic pole of the pulsar. Lower frequencies are generated further out, where the field lines are more divergent. The resonance is due to some combination of magnetic field strength and electron density, both of which fall off with increasing height. So far, so good. But the detail of the resonance and its exact location have eluded us. In fact, the deeper we look into the mechanism of radio emission, the more mysterious it becomes.

Individual radio pulses from any pulsar vary enormously. Improved radio telescopes and receivers, with digital recording, now allow individual pulses to be recorded in detail. Figure 43 shows an example of

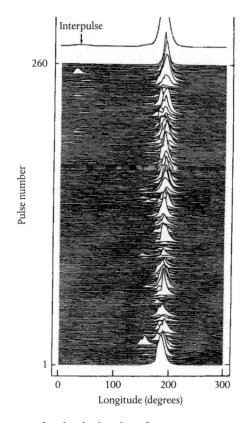

FIGURE 43 A series of individual pulses from PSR B1933+16, showing great variability.

a series of individual pulses recorded using the Lovell telescope. Interpreting this pattern of variation is difficult; it seems that the spread of these pulse components indicates that the source is distributed across a considerable area, so that the variation is a progressive lighting up of different locations within this area. In some pulsars the whole pattern of radio emission disappears for an interval which may last for only a few pulse periods, or in others this null interval may last many hours or days. The most dramatic example of erratic behaviour is the occasional so-called 'giant pulse' emitted by many young pulsars, most notably by the Crab Pulsar. These are obviously much shorter than normal, but measuring their actual duration has been a difficult challenge. Tim Hankins and colleagues eventually used the Arecibo Radio Telescope (the largest radio telescope) and a receiver with a bandwidth of several gigahertz[41] to show that some giant pulses are as short as 2 nanoseconds, which is the time it takes for light to travel 60 centimetres. Apart from the amazing realization that such a short pulse has travelled over 5000 years from the Crab Pulsar without apparent distortion, this observation has profound consequences for theories of radio emission, possibly for every pulsar. Each individual giant pulse must be generated within a very small volume, only about 1 metre across. As Hankins pointed out, this is the smallest object ever detected outside the Solar System, and by far the brightest transient radio source. The phenomenal energy in the pulse appears to be equal to the whole available energy in that small volume of the magnetosphere. We have no idea how to convert energy so efficiently from a static magnetic and electric field into such a short radio pulse. It might be useful to find out.

7

Pulsar Clocks and Relativity

Einstein gave us two theories on Relativity. The first is the Special Theory, which tells us what must happen when we, as stationary observers, look at things and events in some system, such as another planet or a star, which is moving at an appreciable fraction of the velocity of light. The second, the General Theory of Relativity, tells us what must happen when the system is not simply moving but is accelerated, as happens when a planet or star is moving in the gravitational field of another star. Furthermore, the General Theory shows us how to describe these phenomena as the effect of gravity itself on time and space.

Special Relativity is a fairly straightforward geometrical calculation, but General Relativity (GR) is more of a challenge, both to understand and to test. Astronomy naturally provides several tests of GR; the effects of gravity are vital in understanding both the expansion of the Universe as a whole, and the more local effects that a massive star can have on time and space. Cosmology, the description of the largest scales, will be the subject of a later chapter. Here we describe how pulsars have provided the most stringent tests of GR, as it applies in the vicinity of a single massive object, through the fortunate partnership of pulsars with massive stellar companions in binary systems.

The expanding universe, curved space, and relative time are strange ideas. To explain them we usually say: suppose we can find standard measuring rods and standard clocks at large distances in the Universe, or in places where gravity is unusually strong, would they still look the same length and show the same time? One of these 'thought experiments' is close to reality; the pulsars are among the most accurate clocks, they are spread throughout our Galaxy, and some are very close to massive stars. Ideally, a pulsar is an isolated inert and massive body, whose spin should be precisely predictable over very long periods, marking time just like an ordinary but very accurate terrestrial clock. We need to see how well pulsars meet this ideal.

The Fastest Spinners

The main population of pulsars in our Galaxy have rotation speeds between about 10 times per second and once per 10 seconds, i.e. with pulse periods between 100 milliseconds and 10 seconds. Finding pulsars becomes increasingly difficult at shorter periods, and for many years there was no suggestion that any existed with periods much below 100 milliseconds. There was, however, one object in the sky with many of the characteristics of a pulsar: it was small, radio bright, and had the right radio spectrum, but no pulses could be found. Pushing the receiver techniques to detect smaller and smaller periods, it was eventually found in 1982 by Don Backer (1943–2010) to be a pulsar with the astonishingly short period of 1.4 milliseconds. This star, spinning at 641 times per second, was the first of the so-called millisecond pulsars, of which we now know more than 100.

Apart from the amazing spectacle of a star 20 kilometres across rotating at over 600 times per second, the millisecond pulsars have opened up a new era in the measurement of astronomical time. Observing the arrival time of pulses for a few days gives such an accurate description of the rotation that the arrival time of pulses over the next year can be predicted to a few tens of microseconds. This rivals the accuracy of the clocks used as standard timekeepers on Earth; it has even been suggested that pulsars could take over from atomic clocks in defining our universal, worldwide, time. That is not very practical, but instead it has emerged that these clocks in the sky could be used to investigate the fundamental physics of space and time involved in GR.

The millisecond pulsars are often called 'recycled pulsars'. They were originally normal pulsars, neutron stars born in a supernova, spinning at some tens of revolutions per second and slowing down to about 1 per second over a normal lifetime of a few million years. There must be many neutron stars which have slowed down so much that they no longer radiate as pulsars and have effectively disappeared. Some, however, have made a fresh start. Unlike most normal pulsars they had stellar companions, with which they were bound together in a binary system by gravity. The companion stars evolved, as do all stars over a long enough time. Depending on their mass, when a star uses up its hydrogen fuel supply the whole star collapses into a white dwarf,[42] and at the same time hugely expands its atmosphere, so much so that the outer parts can reach its neutron star companion. The strong gravity of the neutron star then sucks the thin outer atmosphere into the neutron star, increasing its mass by a small amount but spinning it up by a large amount. The extra spin comes from the huge angular momentum of the whole binary system, which can spin the neutron star up to rotation speeds approaching a thousand revolutions per second. The limit on speed may be reached when the neutron star itself is blown apart by its own centrifugal forces; the fastest rotation rate yet observed for any pulsar is 716 per second.

Millisecond pulsars rotate more smoothly than normal pulsars, and slow down more slowly. More than 100 are known, all within our

Galaxy. There are some which are solitary, but most are still in binary systems with their original companions which spun them to such high speeds. The companion stars themselves may eventually follow the same evolutionary route, becoming supernovae, and collapsing into neutron stars. These neutron star binaries, consisting of a pulsar and another neutron star in orbit round one another, show the most remarkable behaviour. They are the crucial test beds for relativity theory.

X-ray Binaries

The idea that normal pulsars could be recycled and made into millisecond pulsars may seem impossible to prove, but it turned out that the process had already been observed in another new branch of astronomy. X-ray astronomy cannot be done from the ground, but rocket flights carrying simple X-ray detectors which had been flown in the 1960s revealed the new phenomenon of X-ray binary stars. The first X-ray survey of the whole sky was achieved by the UHURU satellite. Launched in 1970 from San Marco in Kenya, and named after the Swahili word for freedom, this was the first earth-orbiting mission devoted entirely to X-ray astronomy. UHURU detected and located 339 X-ray sources. Among these, one was found pulsating with a period of 4.8 seconds, and it was soon found that the period was varying cyclically through 2.1 days. The same arguments as in the interpretation of radio pulsars showed that this object, Centaurus X-3, was a neutron star rotating once per 4.8 seconds, in orbit round another star in a binary system, with an orbital period of 2.1 days. Unlike the intense radio beams from pulsars, the X-rays were thermal radiation from hot gas or the hot surface of a star. In a binary with the short period of 2.1 days, a companion star must be so close that the neutron star would attract mass from an evolving normal star. The accreting material heats the surface of the neutron star unevenly, so that as it rotates the temperature of the side facing us varies at the rotation rate, and so does the X-ray brightness. For Centaurus X-3, this gives the period of 4.8 seconds; the cyclic variation of this period is a Doppler effect as the neutron star follows its orbit in the binary system.

The companion in this binary was a massive blue star, whose spectrum showed Doppler shifts varying cyclically at the same orbital rate. It happens also that the axis of the binary orbit is nearly perpendicular to our line of sight, so that this massive star actually eclipses the neutron star for about half a day in every orbit. The orbit, and the masses of the two stars, could be described precisely. Furthermore, the spin rate of the neutron star was seen to be increasing, as expected in the recycling process. This was the 'smoking gun' which displayed in remarkable detail the process of creating millisecond pulsars.

Watching—and Correcting—the Clocks

There is a wealth of information to be found by timing the arrival on Earth of radio pulses. Over 700 pulsars are now observed regularly from radio telescopes at Jodrell Bank Observatory and from a network of other radio telescopes around the world. Some have been watched for over 30 years with such precision that every single rotation can be accounted for; in the case of the Crab pulsar, this means that more than 3 billion rotations have been counted since regular timing observations were started. Some pulsars behave so regularly that a sample observation lasting 10 minutes every few weeks or months is sufficient to check their rotation. Individual pulses are usually weak, and an accurate timing can be achieved only by adding thousands together using the best estimate of their period. At each observation the arrival time is compared with an expected time, giving a timing residual that indicates whether the pulsar clock appears to be running fast or slow. These residuals may then be used to detect and measure a binary orbit, or a change in pulsar clock rate, or some more subtle effects.

The timing residuals are sensitive to several factors apart from the pulsar clock itself. Light and radio take 16 minutes to travel from one side of Earth's orbit to the other, so the movement of the Earth round the Sun results in an annual nearly sinusoidal variation of pulse arrival time. This is largest for a pulsar in the plane of Earth's orbit and least for pulsars near the pole of the orbit; the amplitude and phase of the cyclic variation can be used to find the position of a pulsar. A circular orbit is only a first approximation; Earth's orbit is elliptical, as discovered by Kepler. The accuracy of modern pulsar timing observations often approaches 1 microsecond, which is the time for a radio pulse to travel 300 metres. Allowing for the position of the Earth in its orbit demands more than a simple elliptical model of Earth's orbit, since the gravitational pull of other planets has an appreciable effect. The accuracy is already almost sufficient for pulsar observations to make a significant contribution to the measurement of the orbital characteristics of the more massive planets. A smaller, daily variation is due to the rotation of Earth; allowing for this requires an accurate knowledge of the location of the observing radio telescope.

Apart from these geometric factors, there is the effect of travel through hundreds or even thousands of light-years of interstellar space between the pulsar and the Solar System. The interstellar medium is thinly populated with electrons (from ionized hydrogen), which delay the propagation of radio pulses by an amount proportional to the total number of electrons in the line of sight. It also depends in a simple way on the radio frequency at which the timing observations are made; if observations are made at more than one frequency the delay can be found and allowed for.

Assuming all these corrections have been made, the timing observations now relate to the time at which the pulses left the vicinity of the pulsar. The key questions concern the effect of the orbit of a pulsar round a binary companion, but there is first one other geometrical effect to take into account. Many pulsars are moving with high velocities, often some hundreds of kilometres per second. The change of position, known as the 'proper motion' of the pulsar, may be appreciable over some years, or even months, and must be determined and allowed for.

The Relativistic Binary

In 1974 Joe Taylor (b. 1941) and Russell Hulse (b. 1950) started a survey for pulsars using the large reflector radio telescope at Arecibo. Among their 32 new discoveries was a pulsar with a period of 59 milliseconds, which was shorter than any known at the time, apart from the Crab Pulsar. But the period was varying, and after many observations the variation was found to follow the plot of Figure 44.[43]

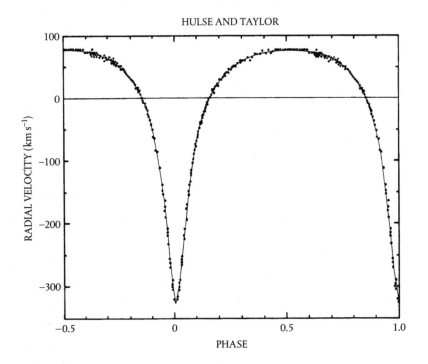

FIGURE 44 Velocity curve for the Hulse–Taylor binary pulsar. *Reproduced by permission of the AAS.*

The period varied cyclically between 58.967 and 59.045 milliseconds every 7.75 hours. The shape of the curve fits exactly the Doppler frequency changes expected from a pulsar following an elliptical orbit around another massive object. In the figure these frequency changes are plotted as velocities, which vary over a range of 400 kilometres per second. A circular orbit would produce a sinusoidal curve; in an elliptical orbit the pulsar moves faster when it is close to its companion, and its velocity changes faster. The lower part of the curve corresponds to a maximum velocity towards us, at this point the pulsar is closest to its companion. The curve defines the orbit, except for a crucial factor: are we seeing the elliptical orbit edge on, or at some angle, possibly even approaching face on? All became clear later, but what was immediately obvious was that in any case the velocity was very high, reaching 0.1 % of the velocity of light, and that the orbit was very small, so small that it would fit into the diameter of our Sun.

Such a close binary was immediately seen as a testing ground for Einstein's theory of General Relativity. GR had already been tested in the planet Mercury's orbit round the Sun. According to Newtonian dynamics, the orientation of this orbit, which is slightly elliptical, should be constant (apart from perturbations by the other planets), but a long series of precise observations had shown that the long axis of the orbit was moving slowly round the orbit. This 'perihelion advance', which amounted to only 43 seconds of arc in a century, was attributed, correctly, to GR. The effect in the pulsar orbit should be much greater, since the pulsar orbit was much closer to its companion, and therefore in a much stronger gravitational field. After only a few years, 'periastron advance' in the Hulse–Taylor binary was found to be a huge 4.2° per year. This was a dramatic confirmation of GR, but more was to come.

Gravity Waves

Newton's revelation that the force of gravity worked over large distances, like the whole of the Solar System, exactly as it does on Earth, did not address a vital question: is the influence of gravity from a distant body instantaneous, or does it travel at a finite speed, which might be the speed of light? This question was answered by Einstein in GR, his General Theory of Relativity. This theory, which shows that gravity does indeed travel as a wave at the speed of light, makes a remarkable prediction about the gravitational effect of a binary system, in which two stars are in orbit round one another. According to Newtonian theory, the gravitational pull of a binary system at a large distance is simply the sum of gravity from the two stars, and should not vary as the stars move in the binary orbit; GR predicts however that there should be a small variation at the orbital period, which should propagate outwards at the speed of light. The effect at a distance is slightly different if the line between the two stars is seen end on or sideways on. This gravitational wave propagates towards an observer, such as ourselves, at the speed of light, so that its effect coincides with our changing view of the binary system. The revolutionary deduction from GR is that this changing gravitational field carries energy away from the binary system, just as an electromagnetic wave carries energy away from an accelerated electric charge. The energy so lost from the binary system comes from the kinetic energy of the two stars. It follows that the energy of the orbital motion must be decreasing, and the orbit must be shrinking. Kinetic energy is being radiated as a gravitational wave.

Gravitational waves are weak. They have no appreciable effect on the orbits of planets in the Solar System, or on normal binary star systems. But the Hulse–Taylor binary is far from normal. The orbit is small enough to fit inside a normal star, and the orbital period is less than a third of a day. It took only 4 years for Taylor and his colleagues to find that the orbit is indeed shrinking at exactly the rate predicted by GR. To great acclaim in the astronomical community, Hulse and Taylor were awarded the 1993 Nobel Prize for this discovery.

The effect of gravitational radiation on the orbit of this relativistic binary is shown with remarkable precision in Figure 45, which covers observations lasting 30 years. The orbital period of 7.75 hours is measured between the times when the pulsar is closest to its companion, i.e. the time of periastron. As the orbit shrinks, this time becomes earlier than expected for an unchanging orbit. The graph shows a progressive

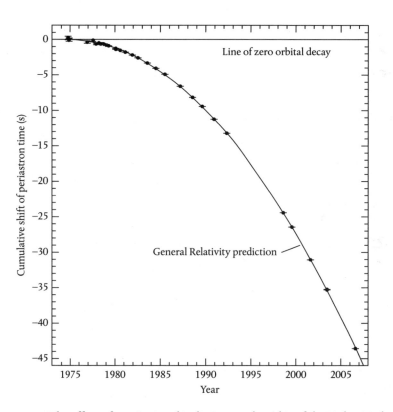

FIGURE 45 The effect of gravitational radiation on the orbit of the Hulse–Taylor binary pulsar. The deviation from constant orbital period is apparent as a cumulative change in orbital phase. *Reproduced by permission of the AAS.*

build-up of the time difference, reaching 35 seconds after 30 years. The points on such a graph should show the accuracy of each observation, but the formal errors in this set of observations are so small that they can scarcely be seen. The rate of orbit shrinking is within a fifth of a percent of the value predicted by GR.

Without the benefit of these precise measurements of relativistic effects, it is impossible to find the individual masses of the two stars in a binary system. The period and size of the orbit are insufficient for this, but adding the two relativistic effects gave very accurate values for the separate masses. In terms of the mass of the Sun as the standard unit, the pulsar mass is 1.4398 ± 0.002, and its companion is 1.3886 ± 0.002. The companion, like the pulsar itself, must be a neutron star, and both masses are very close to the theoretically expected value of 1.4 solar masses.

The Shapiro Delay

The discovery of the Hulse–Taylor binary, in which the pulsar has a period of 59 milliseconds, came before the discovery of millisecond pulsars, which are usually defined as having a period less than 30 milliseconds; most have a period less than 10 milliseconds. Organized hunts for binary millisecond pulsars were successful from 1983 onwards, and it was not long before a millisecond pulsar turned up in a binary system with another neutron star as companion. This offered the possibility of even greater accuracy in verifying GR and even of testing another of its effects. There are now (in 2012) at least seven millisecond binaries with neutron star companions, and many more with white dwarf companions. In some of these the line of sight is such that the pulsar is seen to pass behind, or nearly behind, its companion once in every orbit. In some the extended atmosphere of the white dwarf companion is large enough to eclipse the pulsar, so that a gap occurs in the sequence of radio pulses. The most dramatic effect, however, is seen outside any eclipse, when the pulsar is behind its companion, so that the line of sight passes close to the companion, and through its gravitational field.

As we explain more fully in Chapter 8, Einstein's theory, which applies to any distribution of masses, is very hard to apply except in two cases. Cosmology deals with one case, in which the gravitating mass is uniformly spread throughout the whole universe. The other, which applies to pulsar timing, is the gravitational effect of a single point mass on everything around it, usually described as a distortion of space and time in the local strong gravitational field. This is closely related to the gravitational lens effect, already described in Chapter 5, where we were concerned with the bending of light and radio rays as they pass close to a massive object. The bending of light rays by the gravitational field of the Sun was first observed optically in 1919, when stars close to the Sun could be photographed at the time of a total solar eclipse. The bending of the line of sight to a radio quasar when observed close to the Sun was measured in 1970, when the Sun passed in front of the quasar 3C279. The effect is small, requiring a measuring accuracy of a fraction of an arcsecond. Both the optical and the radio measurements agreed with the predictions of GR within an accuracy of about 10%. Subsequent measurements in the Solar System have improved the accuracy to better than 1%, but tests with far greater accuracy became possible in the binary pulsars.

As well as the bending of radio and light rays as they pass close to a massive body, there is also an effect on propagation time, such that a pulse is delayed as it passes through the gravitational field. This is known as the Shapiro delay, after Irwin I. Shapiro (b. 1929), who proposed a measurement involving planetary radar which would be a test of GR. Radar echoes from planets with solid surfaces, such as Venus and Mercury, can be timed with microsecond accuracy. The planets can be observed by radar with lines of sight that come close to the Sun. For a ray that grazes the surface, the radar pulses are delayed in the Sun's gravitational field by 200 microseconds. The delay falls only slowly with distance, but is negligible when the planet is in the sky opposite to the Sun. The test was proposed by Irwin Shapiro in 1964,[44] and performed successfully on both Venus and Mercury in 1967, and with increasing accuracy on many subsequent occasions. Shapiro was awarded the Albert Einstein Medal in 1994.

Figure 46 shows the effect of this delay, as observed in a millisecond binary system. Through the whole of its orbit, the pulsar is immersed in the gravitational field of its massive neutron star companion. This is the ideal situation, long envisaged by theorists, in which an accurate clock is observed as it traverses a varying gravitational field. The graph

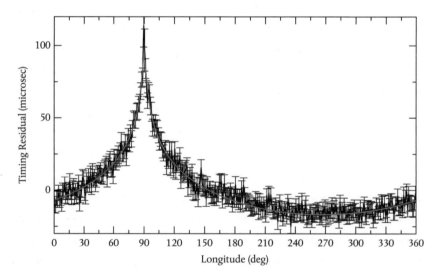

FIGURE 46 The Shapiro delay in the millisecond pulsar PSRJ0737-3039A, which varies over 2.45 hours as it orbits round its companion in the double neutron star system. The peak delay of over 100 microseconds occurs when the line of sight to the pulsar passes closest to the companion, and the travel time of the pulses is most delayed by its gravitational field. (*Michael Kramer, Max Planck Institute for Radioastronomy and Jodrell Bank Observatory.*)

shows how the Shapiro delay varies through the orbit, being greatest when the line of sight is closest to the companion. It is yet another demonstration of the correctness of GR, but it is also very useful in determining the exact configuration of the binary orbit and the precise mass of the companion. In the example of Figure 46, the companion turned out to be another pulsar, in the only known double pulsar system. The Shapiro delay has proved to be an effective determinant of the type of the companion in most millisecond binaries, and especially in picking out those in which the companion is another neutron star.

The Double Pulsar Binary

Having found binary pulsar systems with another neutron star as companion, the observers were competing to find a binary system in which both neutron star components are pulsars. The Parkes survey, which found over 1000 pulsars (see Chapter 6), was the most likely to find such an exotic and exciting object, and all the millisecond pulsar binaries were scrutinized to see if any produced additional pulses at a different rate. The task was shared out among several different research groups, including one at Bologna University led by Marta Burgay. In 2003 she had already found a pulsar with period 23 milliseconds in an orbit with the very short period of 2.4 hours, with a neutron star companion, but with no indication that this companion might be another pulsar. By chance, Duncan Lorimer at Jodrell Bank Observatory was developing a new scheme for detecting binary pulsars, and naturally chose a recording of a known millisecond binary to test his program. Lorimer found the 23-millisecond pulsar, but to his astonishment a much longer periodicity of 2.8 seconds also showed up.[45] Lorimer was lucky. It turned out that this new slow pulsar was only observable for part of the 2.4-hour orbit, and he had chanced on a recording made when its pulses were strong, while Burgay had the bad luck of missing the active part of the orbit. Amid great excitement, Lorimer handed over his result to Burgay, and a detailed investigation began.

The double pulsar binary is an astonishingly rich field for testing relativistic theory. Both periodicities show Doppler shifts as they followed their binary orbits, and the ratio of the Doppler shifts gives a direct measure of the ratio of their masses. But this ratio was already known from the same measurements of orbit precession and shrinking as had been achieved in several other binaries, starting with the Hulse–Taylor binary. This provided yet another test of GR. The two masses, 1.34 and 1.25 solar masses for the millisecond pulsar and its companion respectively, agreed precisely with the earlier calculation. Einstein's GR was giving correct answers to an accuracy of 1%.

The orbit of this double neutron star binary is mildly eccentric. Relativistic precession of this orbit is easily measured, at the phenomenal rate of 17° per year, more than four times greater than that of any other pulsar binary system, and 10^5 times greater than the relativistic precession of Mercury, which had provided the original classic test of GR. Such huge effects prompted a search for yet another phenomenon predicted by GR, 'spin-orbit coupling'. This is closely analogous to an effect in atomic physics, where the spin of an electron is coupled to its orbital motion within an atom. According to Newtonian dynamics, the spin of the individual neutron stars and their orbits in the binary system should be entirely unconnected. According to GR, however, the spin of the individual neutron stars must be defined in the gravitational field of the whole system, so that seen from a distance the spin axis is seen to precess, like the wobble of a spinning top. The effect is that our view of the pulsar is slowly changing, so that the way we see the beam of radiation changes. This appears as a change in the pulse shape, which had already been observed for the Hulse–Taylor binary, but for the double pulsar the effect is much larger; the spin axis of the millisecond pulsar will perform a complete cycle in only 75 years. This is a large effect, which may move the radio beam completely out of our line of sight within only a few years.

The eclipse which confused and delayed the discovery of the slow pulsar in this binary system is a complicated phenomenon.[46] The orbital plane lies very nearly in the line of sight, so that the line of sight to each pulsar passes within 3000 kilometres of its partner. Although the radius of each neutron star is only 10 kilometres, every pulsar is surrounded by a large co-rotating magnetosphere. For the slow pulsar this would normally extend to 132,000 kilometres, well beyond the line of sight to the fast pulsar. There should therefore be a very long eclipse of the fast pulsar, but as observed it is a very much shorter time than expected, corresponding to a magnetosphere diameter of only 9000 kilometres. The radio emission from the slow pulsar is lost for most of the orbit, for a different reason. This is all due to a massive distortion of the magnetosphere, caused by a powerful wind of energetic particles from the fast pulsar. The sketch in Figure 47 shows how this pushes against the magnetosphere, reducing its diameter and creating a long tail which hides the pulses from the slow pulsar.

Apart from its importance in confirming relativistic theory, the double pulsar binary provides a telling example of the recycling process which creates millisecond pulsars. As already described, the spinning up is due to the evolution of a binary companion, which swells up until it spills material onto the pulsar. The pulsar therefore gains both mass and angular momentum, and when the mass transfer is complete it becomes observable as a millisecond pulsar. If the evolution of its companion continues to the stage of collapse to become another neutron star, and this neutron star becomes a normal pulsar, then this will have a period which lies in the normal range from around 100 milliseconds to 5 seconds. This scenario fits perfectly; the millisecond pulsar mass is greater than that of the normal pulsar, due to the mass transfer, and it is spinning more than 100 times faster. It is also a rare specimen among

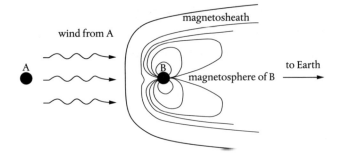

FIGURE 47 Interaction in the double pulsar binary. The distortion of the magnetosphere of the slow pulsar by the pressure of a powerful wind from the fast pulsar A creates a long tail of ionized gas which hides the pulses from the slow pulsar B.

millisecond binaries, which is to be expected since the lifetime of normal pulsars is much smaller than that of millisecond pulsars.

Prospects

Gravitational waves certainly exist; they account precisely for the shrinking of the orbit of the Hulse–Taylor binary, as described earlier in this chapter. Detecting actual gravitational waves as they pass by the Earth is being attempted, so far without success, using large-scale laser interferometers. These instruments should be able to detect short-lived bursts of waves from binary stars in the last stages of the collapse of their orbits, but there are some events on a huge cosmic scale which they cannot detect because they produce only very slow waves. These events are more speculative; a good candidate is the merger of a pair of black holes, which may occur early in the evolution of the Universe, either within a young galaxy or when two young galaxies collide. There may be other, more speculative, events in the early universe. These would produce gravitational waves, with periods of weeks or years. Pulsars clocks may be the only way these ideas could be tested.

Slow gravitational waves can be thought of as a very large scale and slowly changing periodic distortion of space and time. Ideally this could be detected by placing perfect clocks throughout a large volume of the Universe, and observing any slowly changing differences in the time they indicate. An accuracy better than about a tenth of a microsecond, maintained over a period of several years, is needed. Pulsar clocks are nearly up to this specification. A set of 19 millisecond pulsars has already been observed for several years from Parkes, in Australia, and other observatories in the Northern Hemisphere are running similar programmes. Most pulsar clocks already achieve an accuracy of 1 microsecond, so the combination of many such clocks, together with long consistent series of measurements, should achieve the required accuracy. With more clocks and larger radio telescopes, success becomes increasingly likely. This is a prime task for the Square Kilometre Array (Chapter 11).

The next golden aim of pulsar astronomy is to discover a binary in which the companion is a black hole. There is no reason to expect such an exotic object within the scenario of binary evolution outlined above, but it might occur in a more complex interaction within a globular cluster. In some globular clusters there are some dozens of millisecond pulsars, most of them in binary systems, and there is such a concentration of stars that binaries can be disrupted and new partnerships formed during the long lifetime of the cluster. It remains only a speculation, but the promise of new insight into the physics of a black hole is driving the search for more binary pulsar systems. Furthermore, the physics of accretion on to black holes indicates that they might be spinning at rates of some hundreds per second; there should be an interesting effect on the pulsar orbit.

The black hole at the centre of our Galaxy has a mass 4 million times that of a typical neutron star. We see stars in close orbit around it, but the shortest orbital period is 16 years. Suppose we could find a pulsar in orbit round such a huge black hole with a period of only a few hours or days; then we would have a classic clock exploring a gravitational field a million times stronger than the field in which pulsars are already showing the largest relativistic effects so far recorded. A measurement of pulsar precession in such a strong gravitational field would test GR at a very interesting level.

There is a good prospect that a pulsar search of sufficient sensitivity to find at least a thousand binary pulsars will be carried out when the Square Kilometre Array comes into full operation, as scheduled for the year 2024. Among these there may be the most desirable of all binaries, in which a millisecond pulsar is in orbit round a black hole.

8

Radio Expands into Cosmology

Does our Universe go on for ever, both in time and in space? Or did it have a beginning, and will it have an end? Is our Galaxy, the Milky Way, the whole Universe, or is it just one of many galaxies? These questions seemed unanswerable until Edwin Hubble discovered in the 1920s that the Andromeda Nebula was indeed a galaxy like our own but at a greater distance than any Milky Way stars. Using the largest telescope available at that time, the Hooker 100-inch (254-centimetre) telescope on Mount Wilson, he identified individual stars in the Andromeda Nebula, and showed that they were the same type as stars in the Milky Way, but they were so much fainter that they had to be at a distance much greater than any estimate of the size of our Galaxy.

Hubble was building on observations made by several other astronomers who had already speculated on the nature of the spiral nebulae, of which the Andromeda was the best known. Harlow Shapley published in 1918 a model of our Galaxy based on observations of Cepheid variable stars, whose intrinsic luminosity was well established so that their distances could be found from their observed brightness. This gave him a way of assessing the scale of the Galaxy. He also noticed that novae, the very bright but short-lived stars, could be used as distance indicators, and he noticed that novae in Andromeda and similar nebulae were very much fainter that those in our Galaxy. Could they be so much further away that they might be in galaxies like our own, but at huge distances?

In 1926 Hubble published a survey of galaxies, in which he classified them into ellipticals, spirals, and irregulars, a classification which is still in use. Most significantly for our discussion in this chapter, he found that he was looking at galaxies well beyond Andromeda, and he found that the number he could see increased as he looked at fainter and fainter galaxies, according to a simple law. Astronomers measure brightness of stars on a logarithmic scale as *magnitudes*, denoted as m, fainter stars having larger magnitudes.[47] Hubble found that the logarithm of the number of galaxies N increased as $0.6m$, which is what would be expected from a universe which was populated uniformly (on average) with galaxies. On a large scale, and averaging over a large number of galaxies, our Universe was seen to be indeed populated uniformly, with the same average number of galaxies per unit volume at all distances. Hubble even worked out the average density of visible matter in this homogeneous universe, a quantity which is of great significance in cosmology, as we will see later in this chapter.

Best known of all Hubble's work is the relation which he discovered between the velocities of these galaxies and their distances from us. Velocities of stars and galaxies are found from measurements of the wavelengths of spectral lines, which are affected by the Doppler effect. In light from an approaching star all wavelengths are shifted towards the blue end of the spectrum, and from a receding star the wavelengths are shifted towards the red. The fractional wavelength shift measures the velocity as a fraction of the velocity of light.

In 1917, Vesto Slipher (1874–1969) published the results of spectro-scopic observations of spiral galaxies. Using very long photographic exposures, he measured the wavelengths of absorption spectral lines, and found large differences in the wavelength of certain lines as meas-ured in the distant galaxies and in our own Galaxy (and in the labora-tory). The spectral lines were all shifted towards the red end of the spectrum, indicating that the galaxies were all receding from us with very high velocities. Hubble then found that these velocities were pro-portional to the distances of the galaxies, which he had found from the brightness, i.e. the *magnitude*, of the Cepheid variables and other distance indicators. The famous quantity known as Hubble's Con-stant is the ratio of recession velocity to distance. This *redshift–magnitude* relation, as it came to be called, indicated that the whole Universe seemed to be expanding, as though everything in it had once been concentrated in a single infinitely dense point. If the expansion continued for ever, eventually the whole universe would become dis-persed into an empty nothingness. The time scale of this overwhelming behaviour could be deduced from the redshift–magnitude relation; it was about 10 billion (10 thousand million) years.[48] We now had an idea of both the size and the age of the Universe, all achieved using information acquired optically.

Hubble's discovery of the universal expansion, published in 1929, was the main basis of observational cosmology until the early 1950s, when radio astronomy unexpectedly entered the scene. Later in this chapter I describe some theoretical work on cosmology, which I admit, on behalf of my radio colleagues at the time, we scarcely noticed since it seemed not to have any relevance to our new observations. There were also some interesting ideas on the origin of the chemical ele-ments; in particular an explanation was needed for the relative abun-dance of hydrogen and helium, which are the dominant species in the Universe: it was suggested by George Gamov (1904–1968) that this must have been the result of a very dense and very hot early phase of the Universe, like the concentration indicated by looking backwards from the Hubble expansion. (The key paper,[49] published in 1948, was by Alpher, Bethe, and Gamov; the story goes that there were really only two authors, and Bethe only allowed his name to be added for the sake of Greek euphony.) The revolutionary work of Albert Einstein on the structure and evolution of the Universe, which I discuss later, was in the domain of theorists, and outside our radio world. Cosmology was not our business, or so we thought.

We have seen in Chapter 4 that the discoveries of radio waves from the Sun, followed by the discovery of a large number of unidentified discrete radio sources, led radio astronomers to the idea that most of these must be some sort of stars in our Galaxy. Thomas Gold (1920–2004) and Fred Hoyle (1915–2001), who were developing some interesting ideas about the origin of the Universe, became vehement advocates of the opposite point of view, that these discrete sources might be galaxies of some previously unseen type, much further away than Hubble's spiral galaxies, at distances that would be very interesting in cosmology. A famous argument developed in 1951 between the radio astronomer Martin Ryle (1918–1984) and the theorist Fred Hoyle, which was resolved in Hoyle's favour by the identification of the radio galaxy Cygnus A. The large redshift of this galaxy, and of others discovered soon afterwards, showed that most of the hundred or so radio sources were at great distances, certainly outside our Galaxy, and at distances which were significant to cosmologists. Suddenly, cosmology mattered a great deal.

The Steady State Universe

Unfortunately the arguments did not stop there, and developed into a battle between two strong personalities. Hoyle, with Thomas Gold and Hermann Bondi (1919–2005), had pointed out that the expansion observed by Hubble did not necessarily mean that there was a catastrophic start and an eventual fade-away of the Universe. They proposed that the Universe is in a Steady State, in which the dispersion of matter because of the expansion of the Universe was compensated for by the continuous creation of matter out of nothing throughout all space. Only a very low rate of creation was needed to counterbalance the loss observed in the expansion, since the replacement need only occur over the enormous time scale of the apparent age of the Universe. The average density of the Universe is very low; most of the Universe is very nearly a pure vacuum, and spreading the visible material of the galaxies over the whole of space gives a density corresponding to only a few hydrogen atoms per cubic metre. A single hydrogen atom created every year in a volume of several cubic kilometres would be sufficient to replace the galaxies as they disappeared into the remote distance. The arrival of individual new atoms would be such a rare event that there was no way of proving this model by waiting for them to appear in any laboratory apparatus. However, this was a cosmological model which could be tested by comparing distant and local parts of the Universe. In the Steady State there should be no apparent evolution, and on average the number density of galaxies, including radio galaxies, should be the same at all distances. By contrast, in the evolving universe resulting from the Big Bang, the more distant parts must be seen at an earlier age, since light and radio take a time to reach us which, for the most distant galaxies, is comparable to the age of the universe. The earlier, younger universe would have a greater overall density, and presumably a greater density of galaxies.

This crucial test of the Steady State theory would require observations of galaxies at a much greater distance than the visible galaxies which Hubble had been measuring. Radio galaxies were now being found with large redshifts, suggesting that the whole population of radio galaxies might be far enough away for evolution to be observable, either in their individual properties or in the density of their population. The same test that Hubble had applied to counts of galaxies in the local universe could be applied to the more distant radio sources, compiling their number N against the strength of their radio signal, known as their flux density S. By counting visible galaxies, Hubble had found that the local universe was populated fairly uniformly with ordinary spiral galaxies. The number counts of the much more distant radio galaxies should show whether the distant and local regions of the universe had similar populations. Did the local homogeneous Universe extend to cosmologically interesting distances, with much the same density, as it should in the Steady State, or did it look different as it would in an evolving universe with no continuous creation?

I well remember the excitement when this idea dawned on Martin Ryle and his colleagues. Their survey of the sky had unexpectedly yielded a catalogue of almost 2000 radio sources, most of which should be sufficiently distant for the test. The radio astronomers did not follow Hubble exactly in expressing the strength of radio sources in magnitudes, but they did use logarithms of N and S. In the Steady State, and neglecting (as we did at the time) the geometric effects of the expanding universe, there should be a simple relation: $\log N$ must be proportional to minus 1.5 $\log S$. I had already plotted out this relation when we only had $N = 20$ radio sources, a totally inadequate number for the test. Ryle had enough radio sources for a valid test, and a relation which indicated a much larger number of faint sources than expected: the factor in the logarithmic relation was nearer to 3 than 1.5. This meant that the population of radio galaxies was greater in the more distant parts of the universe than locally, which was diametrically opposed to the Steady State model. This appeared to be a decisive test; but were the observations giving a correct view? And was it legitimate to use the simple $\log N / \log S$ test without any consideration of the geometric effects of expansion? This suddenly became a *cause célèbre*, in which Hoyle was criticizing the observational data and Ryle was displaying a disparaging view of theorists in general. The overheated argument even reached the front pages of the national press.

A complication in the argument arose when a similar set of observations was made by Australian radio astronomers, using a different type of radio telescope. Unfortunately their counts of radio sources gave a different result, suggesting that Ryle had over-estimated the numbers of faint sources. As it turned out Ryle's observational data were indeed flawed, but not so badly as to invalidate his conclusion that the Steady State theory was untenable. The argument between Ryle and Hoyle became very public and acrimonious, as described more fully by Simon Mitton in his biography of Fred Hoyle.[50]

The more recent data on the relation between $\log N$ and $\log S$ combines the results of several surveys, in which observations have been extended to sources many orders of magnitude fainter than in Ryle's survey.[51] There are now almost a million known extragalactic radio sources, distributed uniformly over the sky. The observational data presented were obviously entirely at odds with the simple expectation for which $\log N$ is proportional to minus 1.5 $\log S$, when the observations should lie on a horizontal line. The observations have been extended to sources many orders of magnitude fainter than in Ryle's survey; his discovery of an excess number of faint sources was based on a comparatively small number of sources at large values of $\log S$, which do indeed lie above the line. The most obvious feature is a large *deficit* of faint sources. This deficit is the geometric effect of the expansion of the Universe, which becomes overwhelmingly important at the large distances now being reached by sensitive surveys. We need some further theory if we are to interpret the number counts in terms of cosmological models.

What Do We Mean by Distance?

Cosmological distances are usually measured in terms of the redshift z of their light (and radio) wavelength.[52]

For a source receding at velocity v, much less than the speed of light, $z = v/c$; a redshift $z = 0$ means a stationary source. Modern optical observations of individual galaxies may extend to $z = 10$, corresponding to observing them in the distant past.

A direct effect is that the whole radio spectrum is shifted to longer wavelengths, so that what we observe has been radiated by the galaxy at a shorter wavelength (by a factor of two for $z = 1$). Most galaxies radiate less at shorter wavelengths, so that when we see a redshifted spectrum we see a fainter signal. There is, however, also an overwhelming geometric effect of the universal expansion, which we must now explore.

What we see at large redshifts is a universe which has expanded in the same way as the wavelength of light has expanded. This not only changes the way in which the observed intensity of distant objects depends on redshift; at the same time the volume of space corresponding to an interval of redshift no longer follows the local rules. Far fewer sources are present per unit redshift interval, which leads to the fall in the number counts at low flux densities.[53] This geometrical effect applies to all cosmological models.

Geometry is not the only factor which determines the shape of the log N/log S diagram. It turns out that the observed relation is also influenced by the way the universe is evolving. As we look with more sensitive telescopes, we are seeing more distant radio sources; we are also looking back in time and we are exploring the evolution of galaxies a long way back towards the time of their creation. A more critical analysis tells us that the Universe as seen at large redshifts looked very different from the local universe; in particular there is a larger population of radio galaxies and quasars to be observed at distances between redshifts $z = 1$ and $z = 3$ where the radio emission reaches a maximum. Beyond $z = 3$ this population decreases and falls dramatically. The original counts of radio sources showed correctly that Steady State was wrong, but were very far from giving any precise answers to other cosmological questions. Remarkably precise answers were provided later, from radio and infrared observations of the cosmic microwave background, which originated at much larger redshifts; this will be the subject of the next chapter.

Hubble's observations were necessarily imprecise, but he was on the right track about the time scale of the universal expansion which he had discovered. His original observations, however, were still concerned with a small number of galaxies in the local universe. The galaxies available to Hubble and his collaborators in 1929 were limited to redshifts less than 0.01, meaning that their velocities away from us were less than 1% of the velocity of light. We are now able to explore the Universe on a grand scale, looking at the expansion in much more detail. In the rest of this chapter we must now look at the strange geometry of the Universe as a whole, and the way its expansion is controlled by gravity.

Gravity

As I write these words, I am held on my chair by the force of gravity, pulling me towards the centre of the Earth. Isaac Newton showed that gravity is a universal force, not only accounting for the fall to Earth of a ripening apple, but also holding the Moon in orbit round the Earth, and the planets in orbit round the Sun. The same force controls the expansion of the Universe. We start our description of the Universe by extending the Newtonian picture, avoiding at first the complexities introduced by Albert Einstein more than two centuries later.

Consider a simplified universe, in which the distribution of matter is smoothed out, with no concentrations into stars or galaxies. Any part of this universe can then be characterized by its density, the mass per unit volume, which becomes less as the universe expands. According to Newton's law, the matter in every part of this universe attracts the matter in every other part through the same force of gravity that I am feeling as I sit on my chair. The effect, of which Newton was well aware, is that the whole universe tends to collapse in on itself, which is the opposite of the observed universal expansion of the real Universe. But if the real Universe actually started out by expanding rapidly from the energetic forces of the Big Bang, then the force of gravity must be slowing down the expansion, and might even bring it to a standstill and reverse it into a contraction. The dynamics then depend on a balance between the energy of the expansion, and the stored up energy of the gravitational force tending towards contraction. This is the same balance faced by designers of rockets for space research: the pull of the Earth will always slow down the rocket after launch, but depending on the energy of the rocket it will either escape the gravitational trap or fall back to Earth.

In this simple Newtonian approach, the force of gravity must always be tending to shrink the whole universe. This was already known in the nineteenth century, when there was no idea that the Universe was anything other than static; the theory was therefore considered to be wrong, and this simple idea was abandoned. This was unfortunate, since Newtonian cosmology can carry us a long way towards understanding modern cosmology. If we now abandon the idea that the Universe is static, and accept that it is in fact expanding on the largest scale, we can still apply Newtonian theory that there is universal force of gravity which must be tending towards contraction. This means that as time goes on, the expansion must be slowing down. Whether the slow-down is sufficient to eventually reverse the expansion depends on the strength of gravity, which depends only on the averaged-out density of matter. A critical density can be found simply from the Hubble rate of expansion; if the density is greater than this, the expansion will reverse, and if it equals the critical density, the expansion will gradually slow down to zero but never reverse. Now we have to look at the complications which Einstein introduced, which change the equations but not the principle of expansion opposed by gravitational attraction.

Einstein's New Look at Gravity

Einstein gave us two revolutionary principles, already introduced in Chapter 7, which extended dramatically our Newtonian views of dynamics and gravitation. His first, known as the Special Theory of Relativity, tells us what happens when objects move at speeds comparable with the speed of light. Special Relativity has little effect in the everyday world, but it matters a lot in the distant Universe, where galaxies are receding at such high speeds. The second is his General Theory of Relativity, which deals with acceleration, and particularly with acceleration due to gravity. This again has little effect in everyday life, but it has a profound effect on the cosmological discussions of the previous paragraphs.

In Einstein's theory of General Relativity, the effects of gravity and of acceleration are indistinguishable. (This is not so strange when we consider what happens in a lift or a spacecraft, where an acceleration apparently increases or decreases the force of gravity.) He also takes into account the generally accepted fact that an observer in any part of the expanding universe will find that light travels in a straight line, and with the same fundamental velocity. As we have seen in Chapter 7, Einstein's theory leads to a new way of describing the gravitational effect of a single massive body, like the gravitational pull of the Sun on the planets; in addition, gravity appears also to bend rays of light, as we have seen in the phenomenon of gravitational lensing (Chapter 5). Einstein's new look describes these phenomena as a distortion of space, which allows light to travel, as it should, in a straight line but changes the definition of a straight line.[54] He abandons the idea of space as a fixed and immutable frame of reference, like a cosmic piece of graph paper, in which masses may be placed and rays of light shine without any effect on the rectangular coordinates. Instead, the masses create their own frame of reference, within which we must describe the behaviour of moving bodies and rays of light. Close to individual masses this frame of reference is distorted so that moving bodies and rays of light follow the new coordinate system, so that a ray of light can be seen by us as bent, as in a gravitational lens, or the track of a body in orbit round the mass will be affected. What is now needed is a theory of this effect on the largest scale, for a universe with a smoothed out distribution of mass rather than a single concentration.

Einstein's theory sets out how to calculate the effect of gravitation for any arbitrary set of masses, in a set of equations which unfortunately is difficult or impossible to solve except in two simple cases. One of these we have already dealt with in Chapter 7; this is the case of a single massive body, like a star. A more difficult case, but very important, is the binary star system, like the so-called relativistic binary pulsars we encountered in Chapter 7. The other reasonably easily soluble case applies to the smoothed-out universe.

In his General Relativity Theory, as applied to the smoothed-out universe, Einstein adds a fundamental new term to the balance between gravity and expansion. In the Newtonian approach, the attractive force tending towards collapse was due only to the density of matter: in the new theory the local energy density is added to the density of matter. (If it seems odd to treat energy and matter as similar quantities, consider comparing the weight of an empty box in Earth's gravity, and the same box filled with radiation of any kind. You could imagine letting the radiation out through a window, when it would be reasonable to expect the weight to decrease.) Several types of energy are involved in the cosmic equation, including the potential energy created by the material component itself. This is a serious complication, which is the main source of the difficulty in solving Einstein's equations; mathematicians would say it makes the equations non-linear. The solution for the smoothed-out universe is known as the Friedman equation.[55]

The Cosmological Constant

Friedman's equation, which relates the acceleration (or slow down) of the expansion to the density of matter, includes yet another term, which was added by Einstein as an attempt, as he saw it, to make his equations represent a universe closer to reality. He was working at a time, before Hubble's discovery, when there was no reason to think that the universe was expanding or contracting, while his theory inevitably led to collapse under gravity (the same problem faced by the earlier Newtonian ideas). Einstein suggested that contraction was avoided by the addition of a new universal force tending towards expansion. This appeared in his equations, and in Friedman's, as a constant known as the Cosmological Constant. When Hubble showed that the universe was far from static, Einstein withdrew the idea of a Cosmological Constant, and he expressed regret at its introduction. For many years it was thought to be unimportant, and probably even zero, but its reality has now been firmly established by observations of supernovae in very distant galaxies, so firmly indeed that the observers have been awarded a Nobel Prize.

Hubble's galaxy counts, and Ryle's radio galaxy counts, were unable to give anything more than simple information on the expanding Universe. What was needed was a 'standard candle', a type of object that could be seen at great distances, where its apparent luminosity could be compared with the same type of object at small distances. The standard candle was hard to find, but eventually turned up as a special type of supernova, Type Ia, whose origin and brightness suggested that nearly identical objects could be observed at distances out to $z = 1$. At such large distances they would be very faint, and could only be found by using a large telescope dedicated to searching for such newly bright objects lasting for only a few weeks. The 4-metre optical telescope at Cerro Tololo, located in the ideal observing conditions of the Atacama Desert and operated by the Inter-American Observatory, became the main source of newly discovered Type Ia supernovae. The follow-up for each discovery also needed large telescopes, and groups of astronomers all over the world became involved in detailed observations of the spectra and brightness profiles of dozens of faint supernovae. The result was dramatic, both in the discovery of cosmic acceleration and in the manner of its announcement. The two papers which announced the discovery were by two different groups working independently, one by Saul Perlmutter (b. 1959) with 32 co-authors and one by Brian Schmidt (b. 1967) with 23 co-authors.[56]

There was no overlap between the two lists of authors, and fierce competition to claim credit (although the result was so important that having two totally independent derivations was needed anyway). The resulting Nobel Prize, in 2011, was awarded half to Perlmutter and half jointly to Schmidt and his main collaborator Adam Reiss.

The intrinsic brightness of the standard candle turned out to be consistent to within 15%. A comparison between nearby supernovae, with redshifts less than 0.1, and those at large redshifts, between $z = 0.18$ and $z = 0.83$, became possible, and a clear deviation from Hubble's law emerged. In terms of distances determined from apparent luminosity, the distant supernovae were 10–25 % further away than expected from a simply expanding universe with zero cosmological constant. The clear implication was that the rate of expansion of the Universe is accelerating. Independent evidence of the reality of the cosmological constant came later from the observations of the cosmic microwave background, described in the next chapter.

The name 'Cosmological Constant' refers to a term in an equation, but its meaning is usually expressed as 'dark energy'. This is supposed to pervade all of space, and acts as a universal force in the opposite direction to gravity. According to Friedman, this dark energy is another component of the local matter-plus-energy density which determines the effect of gravity on the expansion of the universe; it later turned out, as we will see in the next chapter, that it constitutes no less than 72% of the averaged-out density of the Universe. Only about 5% of the matter-plus-energy density of the smoothed out universe is actual ordinary massive matter (known to physicists as 'baryons'). That leaves 23% for a third component creating gravity, which we have encountered in the description of galaxies in Chapter 5: this is the 'dark matter', whose nature and distribution are both unknown.

Space is flat

If the complications of the Friedman equation have left you gasping for breath, there is some consolation in the title of this section. Recall that in Chapter 7 we encountered Einstein's theory applied to a single material mass in the form of a star, and saw its astonishing success in explaining the bending of light as a distortion, or curvature, of space. Happily we can now regard the space of the smoothed-out large-scale universe as flat! This follows from the observation that the total matter-plus-energy density, which determines the curvature, is exactly (or very closely) equal to a critical density. Euclidian geometry still applies, and we need not worry about bent light rays. But we are left wondering why space is so exactly flat, when the proportions within matter-plus-energy seem to conspire to give a total gravitational effect exactly counterbalancing the Hubble expansion.

An appealing and widely accepted answer to this conundrum was provided in 1981 by Alan Guth (b. 1947). His proposal[57] related to a very early stage in the expansion of the Universe, when instead of the regular linear expansion which we observe today, there would be an enormous exponential growth over a very short time scale. This is known as the 'inflationary model'. It solves two problems. First, any curvature of space is removed; a good analogy is the curved surface of a balloon, which becomes less and less curved as the balloon is inflated. At the same time it accounts for the uniformity of the Universe on the largest scale. How inflation actually occurred is still a matter for speculation, but the concept is generally accepted as the only explanation of the flatness of the present Universe. The event itself cannot possibly be observed; it took place in a time many orders of magnitude less than 1 second after the creation itself, while the most penetrating observations of radiation from the early universe refer to a time 370,000 years afterwards.[58] These observations, which were the next contribution of radio to these overwhelmingly important questions about our Universe, are described in the next chapter.

9

Seeing the Cosmic Fireball

The most amazing and wonderful phenomenon in the Universe is ourselves, the human race, looking at our surroundings and trying to work out where we fit into the whole cosmic scene. We have made astonishing progress in describing our immediate surroundings: the Earth, the Solar System, the Milky Way Galaxy, the expanding population of galaxies, reaching to huge distances which we have learned how to measure but find it hard to comprehend. We also know that it all started with a Big Bang, about 14 billion years ago. We can speculate, with growing confidence, about the events soon after the birth, but it has until recently seemed an impossible dream that we would ever actually see and understand anything that happened in the primitive Universe. This chapter describes how we have achieved exactly that.

In the midst of the noisy controversy between Hoyle and his colleagues, the proponents of Steady State, and the more conventional supporters of Big Bang cosmologies, there were several attempts to understand what actually might have happened in the Big Bang. The most encouraging, in which Hoyle himself was involved, concerned the abundance of the elements. Observations of spectra of light from the stars in our Galaxy had already shown that helium was present universally, with an abundance of 25% compared with hydrogen; it turned out that this was exactly the proportion expected in the synthesis of elements from elementary particles in a very hot, dense fireball. Then there was another prediction: that the fireball would itself be observable, not in its earliest stages but after expansion. In its youngest days and years it cannot be seen at all, and we are limited to observing a surface beyond which no telescope can penetrate; in the same way we cannot see the inside of the Sun, but only its surface, the photosphere. This is a boundary between hot, ionized gas, which cannot be penetrated by radiation, and the later, cooler universe in which ions and electrons have combined, and light and radio can propagate. The boundary is known as the 'last scattering surface'; a better name might be the 'cosmic photosphere'. The stage of evolution at which this occurs is known as the 'epoch of recombination'.

The last scattering surface is the furthest we can hope to penetrate with any of our modern telescopes. It is still hot from the original fireball, but it has cooled to about the same temperature as some hot stars and gas clouds in our Galaxy, about 4000 K. This final barrier to our view of the Big Bang is at a very large distance, so it is in a region of the Universe which is not only very young but is seen at a very large redshift. Our largest telescopes can pick up the faint light from galaxies with redshifts z up to about 10; the redshift at the last scattering surface is about $z = 1500$. Any radiation from the hot gas at the surface would be seen with a spectrum shifted towards long wavelengths by a factor of 1500. Such a huge shift changes the radiation from white light at a wavelength of around 1 micron to short-wavelength radio, at around 1 millimetre wavelength. There is no particular direction where we should look for radiation from the last scattering surface; we as observers are part of the expanding universe, so the radiation from this cooling stage of the Big Bang would be seen all around us.

Herein lies the most important contribution of radio astronomy to the human understanding of our Universe.

Discovery of the Cosmic Microwave Background

In the 1960s, a new measurement of the radio background from the sky was being made at the Bell Telephone Laboratories, the site of Jansky's first radio observations. As in Jansky's work, the intention was to explore the limitations on long-distance radio communications, but at the much shorter wavelengths coming into use for communications between the earth and satellites. The background radio from our Galaxy is mainly at metre wavelengths, and at the chosen wavelength of 7 centimetres this background signal should be very low. A special antenna was built by Edward Ohm in 1961 to measure exactly how low was the background against which satellite radio signals must be detected. The antenna was a 20-foot-diameter horn, like a huge old-fashioned hearing trumpet on its side (Figure 48). The lowest signal Ohm could find was a background corresponding to a temperature of 22 K (degrees absolute), which included unknown contributions from his receiver system and the horn itself. Two radio astronomers, Arnold Penzias (b. 1933) and Robert Wilson (b. 1936), later joined the Bell Laboratories and installed a new receiver with unprecedented sensitivity and low internal noise. They set out to find the origin of the background noise picked up by the horn antenna. One possibility was dirt on the surface of the horn where pigeons had been roosting. Cleaning the whole system, and allowing for the noise generated in their receiver, left a small background contribution of around 3.3 K, which came from everywhere in the sky. By pushing the accuracy of their measurements to the limit, they had stumbled upon the Cosmic Microwave Background.

Not far away in Princeton, and unknown to Penzias and Wilson, Robert Dicke (1916–1997) was building a receiver and antenna

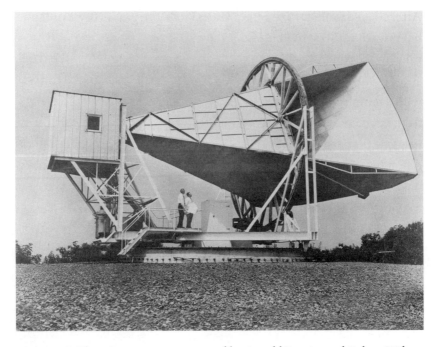

FIGURE 48 The microwave antenna used by Arnold Penzias and Robert Wilson in their discovery of the radio remnant of the Big Bang. *Reprinted with permission of Alcatel-Lucent USA Inc.*

specifically to look for the remnant radiation from the Big Bang, expecting to find a signal of about 3 K. A radio astronomer at MIT, Bernard Burke, who knew what was happening in both research groups, acted as a midwife connecting ideas and observation, and before long the Bell Laboratory group was able to announce the discovery of the Cosmic Microwave Background (the CMB). A Nobel Prize for the discovery was awarded to Penzias and Wilson in 1978.

The CMB is the thermal radiation from the cosmic photosphere, which before the very large redshift would appear to have a temperature of about 4,000 K. The original spectrum of the hot CMB would then be like that of a hot star like our Sun, spanning the visible wavelengths as a smooth spectrum known as a 'blackbody' spectrum.[59]

After the redshift from the early Universe to the present day, the spectrum would still be blackbody, but it would be transformed into radio microwaves and very long wavelength infrared rather than the much shorter wavelengths of light. We would observe blackbody radiation with a temperature reduced by a factor of (1 + z), where z was expected to be around 1000. A measurement of the spectrum was obviously needed as proof of the origin of the radio background, but this was difficult because it requires complete isolation from terrestrial sources of radio noise. This was achieved with spectacular accuracy by the satellite COBE.

The COBE satellite, launched in 1989, carried into orbit at an altitude of 900 kilometres a 'scanning radiometer' FIRAS (Far InfraRed Absolute Spectrophotometer). This measured the infrared temperature over a wide range of millimetre wavelengths. The spectrum it obtained, shown in Figure 49, stands as the best example of any

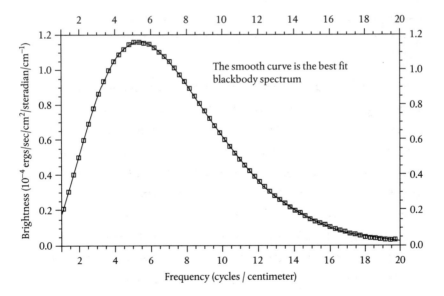

FIGURE 49 The spectrum of the cosmic microwave background, measured by the COBE satellite. *Reproduced by permission of the AAS.*

blackbody spectrum ever measured. The CMB temperature is 2.728 ± 0.002 K measured over the wavelength range 1–10 millimetres. Measuring such a low temperature requires a standard for comparison, so COBE had a large vacuum flask with liquid helium to act as a black body with known temperature. The liquid slowly boiled away, limiting the life of FIRAS to 10 months. During this lifetime the satellite slowly rotated, giving repeated scans of the whole sky which averaged out to give an ever-increasing accuracy. The faint whisper of the Big Bang had been pinned down, and cosmology had entered a new era in which it has become an exact science.

In the world of space research it is difficult to assign names to discoveries; hundreds of people are involved in the years of planning and preparation for the launch of a satellite carrying an infrared receiver and antenna with unprecedented sensitivity. The scientific paper announcing the result had 23 authors.[60] The first author was John Mather (b. 1946), the COBE Project Scientist, and he is credited with the outstanding accuracy achieved in this NASA mission. Ironically, he was nearly upstaged by a small group of rocket scientists in Canada who devised a very similar receiver which was launched within a few weeks of COBE. They measured a blackbody spectrum and a temperature of 2.736 K. If they had managed to do this only a couple of months earlier, we would be celebrating the first measurement of the cosmic background spectrum by Paul Gush,[61] with John Mather's work providing the final accurate touch to the Canadian discovery.

COBE, WMAP, and Planck

The CMB as measured by FIRAS is uniform over the whole sky. But as the primeval gas expanded and cooled, it must at some time have developed structure which evolved into the Universe as we now know it. Can we see any trace of this structure in the CMB? Theory tells us that structure should exist on many scales, which should appear now as a faint mottled pattern covering the whole sky. An action painting by Jackson Pollack provides an example of a large area covered by random blobs, although the CMB would have a very much fainter pattern on a uniform background. The largest scale of the structure in the CMB was expected to have an angular size of about 1°, corresponding to painted blobs 1 or 2 centimetres across seen from a distance of about 1 metre. Another instrument mounted on the COBE satellite, the Differential Microwave Radiometer (DIRBE), was the first to show that the CMB was not perfectly smooth, although it could only distinguish structure on a scale of several degrees. The indication was that the pattern was very faint indeed, only about one part in 100,000 of the 3° radiation. Any measurement has to see this minute pattern within a radio signal that is itself one of the weakest ever to be detected. But it was done by DIRBE, and for this achievement its Project Director, George Smoot (b. 1945), shared the 2006 Nobel Prize with John Mather. The faint pattern certainly existed; what was needed now was a map with better angular resolution.

BOOMERANG is the unforgettable title of the experiment which gave the first detailed picture of the structure in the CMB. It used a high-altitude balloon (Figure 50) to lift a set of microwave radio receivers above the atmosphere, making a long series of repeated scans of the CMB. BOOMERANG was flown almost over the South Pole, with the advantage that the balloon was in a circulating air current of cold polar air which returned it to the launching base after 10 days of continuous operation.

The CMB as seen by BOOMERANG does indeed have a mottled structure with a pattern around 1° across (Plate 10). Here we have a glimpse of the early Universe, like an embryo of a living creature just about to develop the structure of a living adult. The credit for the discovery is shared among an international group of scientists and engineers, 36 of whom are named in the original discovery article.[62]

FIGURE 50 The BOOMERANG balloon launch in 2003. The BOOMERANG radio receivers were carried by this balloon to an altitude of 39 kilometres. *The BOOMERANG collaboration.*

We Are Moving

Apart from small-scale structure, at a size of 1° and less, the cosmic microwave background should be uniform over the whole sky. But what happens when we as observers making a map are moving towards one side of the sky and away from the other? We know that our Galaxy, and all others, have their own individual velocities quite apart from the expansion of the Universe. As with the sound of a siren on a moving motorcar, the Doppler effect shifts the radiation from one side of the sky to shorter wavelengths, and from the other to longer wavelengths, so increasing and decreasing the temperature. The resulting 'dipolar' pattern actually measures our velocity in relation to the Universe. Our velocity is small, only about one-thousandth of the speed of light, so that the Doppler dipole has an amplitude of only one-thousandth of the CMB, but it is easily measureable.

The amplitude of the Doppler dipole is 3.36 millikelvins (mK), which is a fraction 0.0012 of the 2.74 K of the CMB. The velocity of light is 300,000 kilometres per second, so we are moving through the Universe at a velocity of 370 kilometres per second. This is a purely local velocity, which is largely shared by other galaxies in the local group. It is not related in any way to the overall expansion. The dipole pattern has to be removed from all observed maps of the CMB as the first step in analysis, but as an aside it is the only way in which we can define a local standard of rest, against which we can measure velocities in relation to the Universe as a whole.

WMAP: the Wilkinson Microwave Anisotropy Probe

The easily observable spectrum of the CMB stretches from around 20 GHz to 900 GHz; the peak is at 160 GHz. This range, from a wavelength of 15 millimetres to 0.3 millimetres, extends to the shortest wavelengths accessible by radio techniques; in fact the classic black body spectrum obtained by COBE used techniques which are best described as far infrared. The difference in technique is fundamental. Infrared receivers use 'bolometers', which absorb and measure energy, giving a direct measurement of intensity. Radio techniques, in contrast, can deal with the voltage waveform of the incoming signal, preserving both the amplitude and phase. Both techniques were used in COBE, where the first indications of structure were obtained by radio receivers. The next, and most dramatic, radio exploration of the structure of the CMB was made by WMAP, the Wilkinson Microwave Anisotropy Probe.

The wavelength range of the CMB happens to be in a clear gap between two powerful sources of radio waves in our Galaxy. The synchrotron radiation from cosmic ray electrons, first observed by

Jansky, falls dramatically at higher radio frequencies and is nearly negligible at millimetre wavelengths, while at shorter wavelengths measured in microns rather than millimetres the sky is again bright with infrared radiation from warm dust in interstellar space. However, neither of these is completely absent in any map of the CMB, and at the interesting level of only one-thousandth of the genuine cosmic radiation these components of radiation from our Galaxy must be removed. There is also a measurable component of radiation from ionized hydrogen, the 'free-free' radiation described in Chapter 3. These components have different spectra from the CMB, and they may be distinguished by making maps at a range of radio frequencies. The differences between the maps are used to separate out the components with a spectrum of synchrotron, free-free, or dust radiation. The need for this essential correction determined the range of frequencies, 23–94 GHz, to be observed in WMAP.

WMAP measured structure by comparing the strength of the radio signal picked up in two telescopes pointed in opposite directions, mounted in a rotating spacecraft so that they scanned a complete strip of sky. As WMAP slowly orbited round the Sun, this strip scanned round the whole sky, completing a map of structure in 6 months. To achieve this, the spacecraft was launched into an orbit 1.5 million kilometres further from the Sun than Earth, at a point known as Lagrangian L2 (Figure 51). As pointed out by Joseph-Louis Lagrange (1736–1813), the combined gravitational attraction of Earth and Sun keeps a satellite at L2 orbiting the Sun at the same rate as Earth, so that the Sun, Earth, and satellite stay in one line. The WMAP spacecraft stays at the same distance from Earth through the whole 1-year orbit, spinning about an axis aligned with the Earth and Sun. Although WMAP was very distant from Earth and Sun, it was essential to avoid any radiation from either, so the arc followed by the pairs of antennas was kept away from Earth and Sun and protected by a radiation shield.

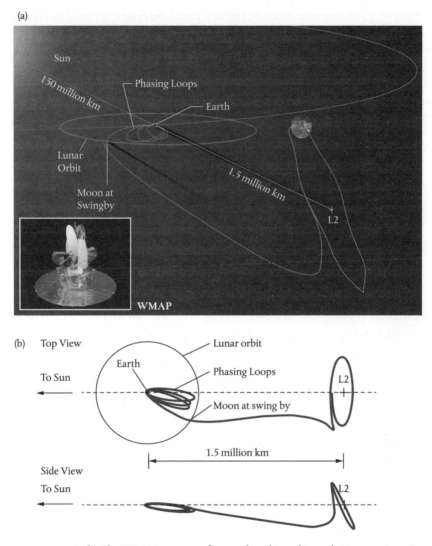

FIGURE 51 (a, b) The WMAP spacecraft was placed in orbit at the Lagrange point L2, as far as possible from the Sun and Earth. The orbit was reached by a slingshot acceleration round the Moon. *(a) Courtesy of Ian Morison/NASA/WMAP Science Team. (b) NASA/WMAP Science Team.*

175

It is cold in outer space, and the receivers settled down to a low temperature only 40° above absolute zero.

By continually and repeatedly scanning the sky, WMAP built up a complete sky map at radio frequencies from 23 to 94 GHz, with an angular resolution of a quarter of a degree (about half the angular size of the Moon). It operated for 9 years, and achieved a sensitivity 45 times better than COBE. In 2003, 2 years after launch, the main results were published in what soon became the most cited scientific paper of the decade.[63] By measuring the size distribution of the mottled structure of the CMB, the WMAP team were able to determine accurate values of a list of cosmological parameters, including the age of the Universe (13.7 billion years), the curvature of space (zero, at 1% accuracy), and the proportions of solid (baryonic) matter, dark matter, and dark energy, together with a possible confirmation of inflation theory and the nucleosynthesis of helium. This was indeed the beginning of precision cosmology.

Planck

After WMAP it might have seemed there was little more to find out about the Universe. But the structure of the CMB had not been fully explored. Both in the far infrared and in microwave radio there were new techniques becoming available which would increase the sensitivity and the angular resolution. These would substantially extend the delineation of detail in the CMB, and also add another parameter, its polarization. But to achieve this, the whole receiving system had to be made much cooler. The Planck spacecraft, launched in 2009 into the same L2 orbit, achieved this with spectacular success. It carried radio receivers with the lowest ever internal noise, and so with the greatest ever sensitivity, at frequencies of 30, 40, and 70 GHz. The frequency coverage continued into the infrared region, with frequencies from 100 to 857 GHz. The infrared receivers were cooled to the remarkably low temperature of 0.1 degree absolute. The sensitivity was so good that structure and polarization could be measured down to a millionth of the total background, so that results are now quoted in units of microkelvins.

Full operation of Planck continued for 2.5 years, ending for the infrared receivers when the helium supply to the on-board coolers ran out. By that time five complete scans had been achieved, and an enormous mass of data had accumulated. Some early results are presented later in this chapter, but before this there are other observations of the CMB to be described.

The Desert and the South Pole

Telescopes carried into orbit by spacecraft are necessarily limited in size, so the aperture of their radio telescopes can only be around 1 metre. As a rough guide, the CMB is brightest between 30 and 300 GHz, or a wavelength range of 10 to 1 millimetres. This means that a 1 metre telescope observing in the centre of the range can have an aperture only around 300 wavelengths across, which limits the detail of any map that it can draw to about a quarter of a degree. This is a serious limitation, which can only be overcome by using larger telescope apertures. Larger telescopes can only be built on the ground, and several are now operating, despite the very serious difficulty that ground-based telescopes have in contending with the effects of the terrestrial atmosphere.

Dry air presents little or no obstacle for radio waves. Water vapour is the real difficulty. Not only does it obscure the cosmic sky over the interesting CMB frequency range, but it occurs in complex random and rapidly changing structures. The worst effects are in a series of resonance bands, but there are frequency bands between the resonances where water vapour has only a limited effect. Radio telescopes working in these gaps and built in a very dry climate can work well enough, and have added significantly to mapping the small scale structure of the CMB. The practical difficulty is that these very dry climates are only to be found in very high or very cold places, which are inhospitable desert locations. Two such sites for CMB radio telescopes are the high deserts of the Chilean Andes, and the high plateau of the South Pole.

Before the era of CMB telescopes, there was a gap between techniques available in the far infrared and the shortest millimetre wavelength radio techniques. The CBM is best observed in this region, and fortunately the gap is now closed by extensions to both regimes. The first measurement of the CMB spectrum, in COBE, used infrared techniques, and WMAP used radio. Planck used both. The same diversity is seen in the ground-based CMB telescopes, and both types have contributed much to our knowledge of the detailed angular structure of the cosmic background.

First, the infrared. A single reflector, several metres across, can have multiple infrared receivers, like those used in Planck, which compare the sky brightness in a cluster of adjacent beams. The whole telescope can then be scanned slowly over the sky, continually recording differences between the separate beams. Such a telescope is SPT, the South Pole Telescope. This is a 10-metre diameter telescope with no fewer than 960 individual receivers in an array at the focus. These can be arranged to operate in various combinations at frequencies of 95, 150, and 220 GHz, so spanning the peak brightness of the CMB. This telescope initially operated from 2007 to 2011, and started full operation in 2012. A similar telescope is ACT, the Atacama Cosmology Telescope, a 10-metre telescope at 5000 metres elevation in Chile; this has a 1024-element receiver array at each of three frequencies, 145, 215, and 280 GHz. These receiver arrays require cooling to the lowest possible temperatures, almost as low as those achieved in Planck.

The alternative approach, which can only be adopted in radio receivers, is interferometry. This is a technique which has already appeared in many places in this book (see Chapter 10 for details). An interferometer can be made sensitive only to small angular structure, giving a direct measurement of exactly what is needed. The required maximum interferometer spacings are no more than the apertures of the infrared telescopes; adequate angular resolution is provided by an overall aperture of no more than 10 metres, so the telescopes look like very small arrays of a dozen or so individual reflectors on a common mounting. The first example of this was indeed called the Very Small Array; it was built at the mountain observatory of the Instituto Canarias de Astrophysica on the island of Tenerife. This was followed by CBI, the Cosmic Background Imager, located high in the Chilean Andes. This is a 13-element interferometer mounted on a 6-metre platform giving maps with a resolution of 4.5 to 10 minutes of arc at radio frequencies between 26 and 36 GHz.

Between them these adventurous new telescopes have revealed the fine structure of the CMB; we now turn to its interpretation.

Measuring the Ripples

Like ripples left in a sandy shore when the tide goes out, the mottled structure of the CMB has a characteristic size. On a flat surface we could describe the pattern in a graph of ripple amplitude against ripple width. On the spherical sky the ripple width becomes an angle, which is described according to the way in which the structure fits into the spherical sky; mathematically the width of the ripples is expressed in terms of 'spherical harmonics'. The lowest numbered harmonic describes a simple dipole, like the dipolar pattern due to the motion of the Earth (We are Moving), which is numbered $l = 1$. The smaller the structure, the larger is the l number.[64]

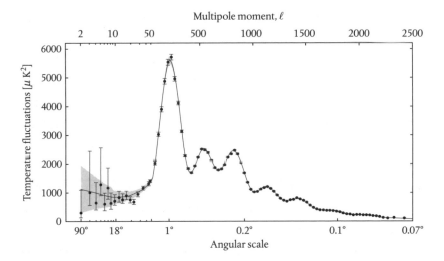

FIGURE 52 The structure of the temperature ripples in the cosmic microwave background, presented as an angular spectrum. Spherical harmonics are characterized by their l=number. A spherical harmonic scale is shown above. *ESA and the Planck Collaboration / Planck Collaboration, 'Planck 2013 results. I. Overview of products and scientific results' March 22, 2013 issue of Astronomy and Astrophysics.*

The whole pattern specified in this way is shown in Figure 52. This is the famous spectrum of the CMB structure, made initially by the WMAP spacecraft, and seen in more detail by Planck.

The spectrum is full of information, from the position of the main peak at $l = 180$, corresponding to ripples on a scale of 1°, to the harmonic peaks at the larger l numbers. The full line shows the astonishingly good fit with a theoretical model.

The Rich Field of the Cosmic Microwave Background

The steps in observational cosmology, from the discovery of the CMB through to the measurement of its fine detail in the spectrum of Figure 52 represent a tour de force without precedent in the history of science. So too does the theoretical work which has explained every detail of the spectrum and in so doing has turned cosmology from arcane speculation into an exact science. The theory is complex: the following account is an introduction rather than an exposition.

The prime reason for measuring the structure of the CMB was to match it to the structure of the visible Universe. The galaxies, and especially the groups of galaxies, observable at great distances, must originate in the ripples which we now see with a scale of several arcminutes in the CMB. But the CMB depicts the Universe as it was when it was about a thousand times smaller than it is now; we know this because the wavelength of the CMB is about a thousand times longer than it was when it was radiated. When this huge correction of scale is made, the fit to the Universe as we know it is remarkably good. Incidentally the success of this comparatively simple piece of geometry tells us that no correction is needed for a possible curvature of space; it is a comfort to know that the curvature must be zero, or very nearly so.

The more complicated bits of the theory concern what must have happened inside the cosmic photosphere to produce the ripples in the first place. The spectacular image of the CMB produced by WMAP and Planck represents a slice through the pregalactic Universe at a time 370,000 years after the initial Big Bang. Although we cannot see the earlier Universe beyond the cosmic photosphere, we can make a remarkably good model which takes us back to an age of less than a second. At that early stage of the expanding fireball there had already been two important events; some of the original hydrogen had been combined into helium, and the whole Universe had undergone a huge expansion, known as the inflation. The development of structure in the fireball, from an age of only 1 second to 370,000 years, can be followed in remarkable detail. The theory is like that of sound waves, but adapted to an expanding medium. Small fluctuations in density grow, due to gravity, but with the principles of General Relativity reminding us that energy in radiation adds to the effect of mass in the force of gravitation.

Up to this point in the discussion, the description of the early Universe was good enough to work out the scale and amplitude of ripples that should be generated in it. It turned out, however, that even if the model included several different components of radiation, and matter consisting of ionized hydrogen and helium, it required another more mysterious component to make it fit the observed structure. This component, which increases the effects of gravitation, is the 'dark matter' which has already turned up in our discussion of the dynamics of galaxy rotation (Chapter 3). Another unfamiliar component of the Universe, which becomes important long after the epoch of recombination, is 'dark energy', which we have encountered earlier in a discussion of the Cosmological Constant (Chapter 8).

Making Ripples

The spectrum of the ripples in Figure 52 resembles the spectrum of sound from a bell, with a dominant fundamental note that is determined by the size of the bell, and several harmonics which are related to the shape of the bell and the way it is struck. The ripples have a dominant size of around 1°, and harmonics at nearly arithmetic ratios around ½°, ⅓°, etc. Every detail of the spectrum is significant, especially the size of the peaks and the curve to the left of the main peak, which relates to structure on the largest scale and a possible detection of the effects of inflation.

Two concepts dominate the physics of the fireball, both before the epoch of recombination and up to the present day. First, the universal tendency of gravity to form condensations on any scale; second, the universal expansion, which tends to drive apart any structure as it develops. There is a sharp change in the way these work at the epoch of recombination. What we see in the CMB is the structure which has developed up to recombination, observed as temperature changes in the ionized matter. Before recombination the three components, ionized gas, radiation, and dark matter act together in the developing structure, while afterwards the structure in only the dark matter and the gas develops independently into the galaxies and groups of galaxies we see today. The way in which the structure develops up to recombination depends on the mixture of the three components and the time available for condensations to develop. Since we know the age of the Universe at recombination, we can relate the observed spectrum of ripples to the proportions of the three constituents; this allows us to make some surprising, and surprisingly accurate, statements about what went on in the first 370,000 years.

The theory of gravitational collapse in a static gas cloud was formulated in 1902 by James Jeans (1877–1946). He showed that condensation beyond a minimum size would increase indefinitely. The theory was adapted for an expanding universe by Evgenii Lifshitz (1915–1985), who provided a more general solution, showing how ripples develop with time. The discovery of the CMB in 1964 stimulated the Moscow cosmologists Yakov Zel'dovich (1914–1987) and Rashid Sunyaev (b. 1943), and James Peebles (b. 1935) at Princeton, to develop the theory, showing how the growing condensations developed up to the epoch of recombination, and then became frozen into the spectrum depicted in Figure 52. The physical situation from an age less than 1 second right through to 370,000 years can be understood in classical physical theory, with the additional insight from the special and general theories of relativity that the energy in radiation has a gravitational effect. Radiation in fact initially dominates the scene, with the concentrations of ionized gas following concentrations in radiation energy.

Condensations which are large enough to grow are unstable, and continue to grow. Below a certain physical scale, known technically as the Jeans length, the perturbations are stabilized by the internal pressure of the photon-dominated plasma and they oscillate at a rate depending on the density of matter and energy within them. Condensations on the largest scale undergo only one oscillation before the epoch of recombination; these form the dominant 1° pattern in the ripples. Smaller condensations appear after two or more oscillations, giving the harmonic series in the spectrum. The details of the spectrum depend in subtle ways on the proportion of inertia provided by the mass in the later stages of expansion, but it is the 1° scale and the amplitude of the first peak that show unequivocally that the ionized hydrogen and helium do not provide sufficient gravitation for the process of condensation.

This revelation proved that the observable matter (hydrogen and helium, often referred to as baryonic matter) can only constitute a small proportion of the material Universe. Dark matter, whose only apparent function is to provide gravitation, is five times as important. A second revelation comes from the observed angular scale of the ripples, which shows that the curvature of space is zero. But this reassuring conclusion leads to another question, since the curvature of space depends on the density of matter and energy which it contains. A sufficient density is only obtained by adding the final component, dark energy. This plays little part in the formation of structure in the early universe, but it dominates the expansion of the Universe on a large scale at very late epochs, at redshifts $z < 1$.

The precision of the observations, and the straightforward nature of the theory, result in precise values for the proportions of the three components. The Universe as measured by Planck contains 4.9% baryonic matter, 27% dark matter, 68% dark energy. More than 95% of the Universe is invisible, and the stars and galaxies we have been studying so assiduously make up less than 5%. There is plenty of scope for further understanding this strange situation.

After Recombination

The map of the CMB and the spectrum of its structure have been presented in this chapter as though they were made from a single map of the whole sky. In practice, maps were made by WMAP and Planck at several frequencies spread from 20 GHz to 900 GHz (wavelengths 1.5 cm to 0.3 mm), so that small but significant components of radiation from the Milky Way galaxy and more distant galaxies could be evaluated and subtracted. These components have different spectra; when they have been sorted out they prove to be interesting in their own right. At the short wavelength end of the CMB the map of the sky is dominated by radiation from clouds of dust and molecules which pervade much of interstellar space in our Galaxy; Plate 11 from Planck shows these clouds stretching far beyond the concentration of stars in the visible Milky Way. Plate 12 shows the CMB when the local components have been subtracted.

Some individual distant galaxies can also be seen as individual radio sources, although radio from most of these has such a steeply falling spectrum that they are not easily seen at such high frequencies. There is however an effect related to clusters of galaxies seen against the background of the CMB. This is the Sunyaev–Zeldovich effect, which occurs when the CMB is observed through the cloud of hot gas which occurs within such clusters. The electrons of the hot gas interact with the photons of the CMB, cooling the electrons and increasing the photon energy by a small amount. The whole blackbody spectrum can be shifted by this effect, which reduces the brightness at the lower radio frequencies, and has been observed by Planck in a number of clusters of galaxies. These effects, which may initially be regarded as incidental to the basic aims of WMAP and Planck, are proving to be rich fields of research, and are already the subjects of many research papers.

The history of our Universe after the epoch of recombination, and before the formation of the galaxies and cluster of galaxies that we can observe today, is still largely unknown. The tiny ripples which have been measured so precisely in the CMB must at some stage build into larger and denser concentrations, where stars would eventually be born. After recombination, the material Universe was mainly composed of neutral hydrogen and helium, and the developing structure must appear as structure which we might be able to observe. This requires a new regime in radio astronomy, observing the 21-centimetre hydrogen line at an appropriate redshift, probably at a wavelength increased by a factor of around 10 or more. A new generation of ground-based telescopes is now being constructed which should achieve this; these long wavelength radio telescopes are technically very challenging, as described in the next two chapters.

10

Big Dishes and Arrays

When I started research with Martin Ryle in 1947, we had no idea that our simple radio observations could be developed into the wonderful contributions that continue to be made by radio astronomy today. Radio has added to our understanding of the Universe on every scale, from the compact neutron stars to the expanding cosmos itself. Radio uses wavelengths which are around a million times longer than light waves, and it seemed impossible that maps could be made of any radio object with a precision comparable with optical photographs. Three technological developments have unlocked the mapmaking potential of radio astronomy: increasingly sensitive radio receivers, digital computers, and above all the use of arrays of interconnected radio telescopes. Arrays, which act as huge single telescopes, are only possible because of our ability to detect and manipulate the oscillating electric field in a radio wave, a technique which is almost impossible for light waves. But before we can build an array, we need to collect as large a signal as possible by building large individual reflector telescopes, the 'Big Dishes', and using the most sensitive radio receivers.

The Big Dishes used by radio astronomers can be compared with large optical telescopes. In 1947 the largest of these was the 200-inch

(5.1 metre) Hale Telescope coming into use at Mount Palomar. It is still in use, and it was the largest for over 40 years. There are now several optical telescopes with 10 metres diameter, and plans to build giants with diameters up to 40 or 50 metres. This push towards large size applies to telescopes for any wavelength, from radio to gamma-rays: a larger aperture collects more light, and it can make sharper images. Radio has followed the same path, with telescopes up to 100 metres in diameter, which are good for picking up very faint radio signals but still not good enough for making maps. Sharp radio images can only be obtained by using apertures which are many thousands of wavelengths across, which for typical radio wavelengths means apertures measured in kilometres rather than metres. In this chapter we describe the giant radio reflectors, and show how their limitation in making sharp images has been overcome by interconnected arrays.

It was the need for a large collecting area that prompted Bernard Lovell to design the first large steerable reflector telescope at Jodrell Bank. The intention was to explore new sources of radio waves, from the Milky Way and other galaxies, and possibly to detect radar echoes from cosmic ray showers. The original design was for wavelengths of 1 metre or longer, which required the reflector surface to be accurate to about 10 centimetres. This accuracy could be achieved for the large diameter of 250 feet (78 metres). But a problem arose during design. A very important new source of radio, hydrogen line radiation, was discovered in the Milky Way galaxy, which required the telescope to work on the much shorter wavelength of 21 centimetres. A more accurate surface was included in the design, at an extra cost which nearly halted the whole project, but which gave us the famous telescope we know today as the Lovell Radio Telescope.[65] Since its first operation in 1957 the Lovell Telescope has had two major improvements, including a new reflecting surface in 2002, and it now operates efficiently on wavelengths as short as 5 centimetres. At this wavelength the beamwidth is about 3 minutes of arc, which is nearly twice as good as the visual acuity of the human eye.

The best that can be achieved for a reflector radio telescope that can be steered and pointed to any part of the sky is represented by the 100-metre-diameter GBT (Green Bank Telescope) in West Virginia, USA. This has a reflecting surface with sufficient accuracy to work at wavelengths as short as a few millimetres, giving a beam about 10 arcseconds across. A larger but less precise parabolic reflector, 1000 feet (305 metres) in diameter, was built at Arecibo, Puerto Rico, in 1963. This reflector telescope (Plate 15) is fixed to the ground and can only be pointed upwards, in directions near the zenith. Its beam can be steered to angles up to 20° from the zenith by moving the detector system at the focus; this is carried on a massive structure, weighing 900 tons, suspended 125 metres above the reflector surface.

An even larger reflector, the Five hundred metre Aperture Spherical Telescope (FAST) is being built in Guizhou province, southwest China. Like the Arecibo reflector, FAST is built in a large natural bowl in the limestone terrain known as karst. This telescope is probably the largest single parabolic reflector telescope that can ever be built. The reflector surface is made up of 4600 triangular panels, whose individual positions can be adjusted to give sufficient accuracy for wavelengths as short as 6 centimetres. By adjusting the shape of the surface as the detector system is moved at the focus, the telescope beam can be moved at angles up to 40° from the zenith, although only 300 metres of the surface can then be used. At the shortest wavelength of 6 centimetres, the beamwidth will be less than a minute of arc.

FAST will be uniquely useful as the most sensitive for detecting and monitoring weak radio signals from individual objects such as pulsars and space probes, and in the search for signals from intelligent life in distant planets. But it will not on its own have a sufficiently narrow beam to produce detailed maps of any astronomical objects. For this we require angular resolution better than a second of arc, or even a thousand times better, at a milliarcsecond. Surprisingly this can be achieved at radio wavelengths more easily than at the much shorter wavelengths of optical telescopes.

Interferometers

Angular resolution has always been a key concept in astronomical observations. The sharpness of images in any telescope is fundamentally limited by the diameter of its aperture; the angular resolution can be no better than the ratio of wavelength to the diameter of the aperture.[66]

An optical telescope with aperture only about 20 centimetres across can in principle obtain images smaller than one arcsecond across, although telescopes with larger apertures may not have a proportionate improvement in image size, because images may be blurred by atmospheric scintillation. The Hubble Space Telescope, with an aperture of 2.4 metres, is clear of the atmosphere and achieves an angular resolution better than 0.1 arcsecond. At radio wavelengths, which are longer than light waves by a factor of around a million, even the aperture of the parabolic telescopes shown in Plates 13, 14, and 15 is typically only several hundred wavelengths across, giving an angular resolution of around 0.1°. Nevertheless modern radio telescopes do achieve angular resolutions as good as the Hubble, and make detailed maps at such high resolution. This is achieved by *aperture synthesis*, a revolutionary process which is fundamental to a generation of large radio telescopes.

A radio telescope constructed by aperture synthesis may be many kilometres across, or even some thousands of kilometres across, but the only parts of this huge aperture which actually exist are pairs of much smaller elementary telescopes connected together as interferometers. The idea is shown in Figure 53. Here a conventional parabolic reflector is seen to collect radiation from a distant source and send it to a focus. Dissecting the aperture into small areas, each with its own receiver, and sending the separate signals to a single receiver by transmission line, can achieve the same result. All that is required is to keep the signals in step by using identical lengths of transmission line. The trick of aperture synthesis is to treat these signals one pair at a time, so that the whole aperture can be built up from a series of interferometer pairs, each of which might exist for only a short time.

The earliest steps in radio interferometry, which were described in Chapter 2, used only two separate antennas. The two-element

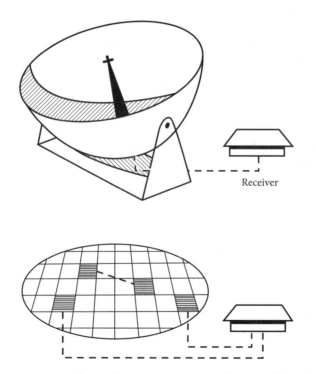

FIGURE 53 Dissecting a large telescope aperture, and re-assembling the pieces.

interferometer used by Martin Ryle, and the sea-reflection interferometer used by Joe Pawsey, both in 1947, were intended to analyse the location and diameter of the radio sources in the Sun. As described in Chapter 2, the original idea was to connect two small antenna arrays, spaced many wavelengths apart, to the input of a single receiver. The combined reception pattern was the usual wide beam of a single array, but divided into lobes separated by an angle which was the ratio of the wavelength to the spacing; for example, at a spacing of 100 wavelengths the lobes would be 0.01 radians apart, i.e. just over half a degree. It turned out that the radio source in the quiet Sun was larger than this, while the enhanced sunspot source was smaller. As shown in Figure 10, the response as the Sun moved across the sky was a sinusoid from the small source together with a smooth curve from the whole source. This was a way of analysing the brightness distribution across the Sun.[67] There was still a big step, which we will explore in this chapter, from analysis to synthesis, but it is clear that the basic principles were already understood by several research groups by the early 1950s.

Following the discovery of the strong radio sources Cassiopeia A, Cygnus A, and Taurus A, the research groups in Australia, and at Jodrell Bank and Cambridge in England, set out to measure their angular diameters. Were they star-like, or were they much larger, like the Crab Nebula which was already tentatively identified with Taurus A? No single antenna could achieve the necessary angular resolution, and an interferometer had to be used. Furthermore, it seemed that interferometers with large spacings of many wavelengths would be needed, so that connecting the two antennas to one receiver could not be achieved by conventional coaxial cables. Radio links joining an outstation several kilometres away to a home telescope were devised in Australia and at Jodrell Bank. In Cambridge we already had a much simpler cable-linked interferometer (which was my PhD project), but it had a rather short baseline. Fortunately for me and my colleagues it turned out that the radio sources were larger than expected, and my interferometer was long enough to achieve the necessary resolution to measure the angular diameters of both Cygnus A and Cassiopeia A (see Chapter 5). Soon afterwards, the radio link technique allowed Roger Jennison (1922–2006) and Mrindal Das Gupta (b. 1923) at Jodrell Bank to use longer interferometer baselines and improve on the simple measurement of diameter: they were able to show that Cygnus A was a double radio source.

The radio link technique was then applied at Jodrell Bank in the 1960s to measure the angular diameters of the increasingly large numbers of radio sources that were revealed in the Cambridge surveys. Large spacings were necessary, and a transportable antenna was eventually used to set up an interferometer with a 134-kilometre baseline stretching across England, between Cheshire and Yorkshire. Even at that large spacing some sources remained unresolved, with diameters that might be as small as those of stars rather than galaxies; these were therefore named the quasi-stellar sources, or quasars. Other sources, later identified as radio galaxies, showed interesting structures, like the double structure of Cygnus A. But no detailed maps could be drawn; that would require the next stage, which became known as aperture synthesis

Aperture Synthesis

To understand aperture synthesis, we need to make a small diversion into the ideas of Fourier theory. Fourier theory is concerned with any variable quantity, like the air pressure in a sound wave. If a pipe organ plays a chord, the sound has a spectrum with peaks at the frequencies produced by each of the pipes being sounded, together with their harmonics. The complex sound is a blend of components, each a pure tone with its own frequency. Similarly a distribution of radio brightness across an object in the sky can be analysed as the sum of sinusoidal components in space, rather than in time, each with its own spatial wavelength. The analogy is particularly close for an impulsive sound, like the beat of a drum or the clash of cymbals, whose explosive sound carries no predominant frequency but a continuous spectrum covering much of the audible range. A celestial radio source might similarly be a single isolated object, whose analysis in space would be a continuous spectrum of sinusoidal components coinciding and adding at the centre of the radio source. In both cases it would be possible to synthesize the impulse, or the discrete source, by adding enough sinusoidal components arranged to add at the right moment in time, or at the right position in space. This is the process of Fourier synthesis.

If we now put together the idea of building up a map of a radio source by measuring separately the sinusoidal components of its brightness distribution, and the early concept of an interferometer measuring one such component, we reach the basic concept of aperture synthesis. If the radio source is observed repeatedly with an interferometer with different spacings, the outputs of the receiver can be added together to make a map of the distribution of radio brightness across the source. Developing this idea was the main theme of the Cambridge radio observatory over a decade, during which a series of telescopes were built, culminating in the first recognizably modern synthesis telescope, the Cambridge One-Mile Telescope.

Suppose we want to build a telescope aperture some hundreds of wavelengths across, using only two small elementary units each several wavelengths across. Figure 53 shows a chequer-board representing the whole aperture, with the two elements blocked in. These two are used as an interferometer, and as the sky drifts across the wide beam of the small aperture we can record one Fourier component of the radio brightness. The receiver outputs must be recorded and added, giving the full two-dimensional map of the part of the sky within the beam of the small unit aperture.

A less tedious scheme is to construct one of the units as a long array extending along one side of a square, which is conveniently aligned east–west, with a single smaller antenna which can be placed successively at stations along a north–south line. This 'moving-tee' interferometer system was the first aperture synthesis system at Cambridge, built by John Blythe in 1957. I built a larger version later, with Canadian student Carman Costain (1932–1989); this had an east–west arm 1 kilometre long. Using a wavelength of 8 metres, this gave maps with a beamwidth of 0.8° × 0.8°, an unprecedented resolution for such a long wavelength.

A related design was the Mills Cross, built in 1954 and shown in Figure 54. It consisted of two long elements, respectively east–west and north–south. In a later version each of these was a cylindrical paraboloid, 1600 metres long and 12 metres wide. Using the shorter

FIGURE 54 The Mills Cross radio telescope at Molonglo. A later version using large cylindrical paraboloids is still in use. *CSIRO Radio Astronomy Image Archive.*

wavelength of 72 centimetres, this produced a single beam only 3 arc-minutes across; there was no need for repeated observations, as all interferometer spacings are present in this configuration. The east–west arm of the Mills Cross was later used in an important survey of the Southern Hemisphere for pulsars, and it is still in use as a test-bed for the receiver systems of the Very Large Array (Chapter 11).

The Mills Cross was the last of this design; Martin Ryle was already working on a new scheme called 'Earth-rotation synthesis', in which a two-element interferometer need only be arranged on a single east-west line. If observations and recording were continued for 24 hours, the rotation of the Earth swept up a track across the desired telescope aperture. Moving one of the elements by its own width each day was all that was needed to cover the chequer-board of Figure 53.

Looking back from this computer age, it is hard to realize that at the time of these developments there were severe technical problems to be overcome. The main difficulty was recording the output data, and subsequently feeding it into a computer to synthesize the map. Miles of punched paper tape were used for recording; there was no other form of digital recording. Computers were in their infancy; the newly invented EDSAC II computer in Cambridge was the first to be used for processing the data. Despite the difficulties, the first 'Earth rotation synthesis' map was made by Martin Ryle and Ann Neville in 1962.[68] This was a survey of a region of sky round the North Pole, so that the elementary antennas did not have to move to keep the survey region in the centre of their beam for 24 hours. Ryle then constructed his famous One-Mile Telescope (Figure 55), which could track any region of the sky, making the first high resolution maps of Cygnus A and Cassiopeia A (at a resolution of 23 arcseconds: see Figure 31) among many others.

Aperture synthesis was evidently the way forward. One further step was needed: was it necessary to fill all the spacings on the chequer-board, or along the east–west line of the Earth-rotation instrument? Leaving

FIGURE 55 Martin Ryle's One-Mile Synthesis Radio Telescope at Cambridge. © *Neil Grant/Alamy.*

out some spacings has an interesting effect on the maps; it leaves the images just as sharp, but they are surrounded by a spurious pattern known as sidelobes. These can be removed in the data analysis, but only at the cost of extensive computations. As at every stage of this story, putting this into practice depended on the rapidly improving capabilities of digital computing. This is still the case; the Square Kilometre Array, which is being designed to be fully operational in around 2025, will require computer power considerably greater than anything available today.

The successful demonstration of aperture synthesis at Cambridge was immediately appreciated by radio astronomers worldwide. At the time there was a proposal for three European countries, Belgium, Netherlands, and Luxembourg, to join in building a new radio telescope. Professor Jan Oort, of the Netherlands, who incidentally was one of the few astronomers from traditional optical fields who saw the possibilities of radio astronomy, was leading this collaboration. When he saw the map of the North Pole region produced by Ryle and Neville, he changed the whole direction of the so-called Benelux proposal, and started the design of a new synthesis array. The collaboration eventually came to nothing but the array was built nevertheless, at Westerbork in the Netherlands, and became one of the most important radio telescopes in the world (Figure 56).

The Westerbork Synthesis Radio Telescope (WSRT) consists of 14 parabolic reflectors, each 25 metres in diameter, on an east–west line 2.7 kilometres long. Four of these are on rail tracks, so that all spacings up to 2.7 kilometres can be used. Interchangeable receiver systems cover a wide wavelength range, from 3.6 centimetres to 2.7 metres. Remarkably for a telescope which first operated in 1970, the WSRT is

FIGURE 56 The Westerbork Synthesis Radio Telescope (WSRT). Each of the 14 parabolic reflectors is 25 metres in diameter. © *Chris Sciacca, IBM Research.*

still at the forefront of technology. It was upgraded in 1995–2000, and further improvements are planned. It can be used in a simplified configuration, in which all individual telescopes are combined to produce a single large collecting area, making a single telescope with a sensitivity similar to that of the largest single parabolic telescopes. This is particularly useful in very accurate timing observations of pulsars, when it forms part of the European Pulsar Timing Array.

Aperture synthesis was adopted in the USA in a spectacular new radio telescope, the Very Large Array (VLA), built on the plains of St Augustin near Socorro, New Mexico, and first operated in 1980 (Figure 57). The location was chosen because it is an area of low radio interference, but also because of the sheer extent of the instrument: 27 parabolic antennas are arranged in a Y formation, with maximum spacing no less than 36 kilometres. They are all mounted on rail

FIGURE 57 VLA, the Very Large Array. Twenty-seven parabolic reflector antennas are arranged in a Y formation. The maximum spacing between antennas is 36 kilometres. *Image courtesy of NRAO/AUI.*

tracks, which allow them to be placed either in a very open array, or in a more compact array all within a maximum interferometer spacing of 1 kilometre. The open array is often referred to as a sparsely filled aperture, and the more compact array is often used to overcome the complications of missing Fourier components. Interconnections originally involved waveguides, but these have now been replaced by fibre optic cables. The wavelengths in use range from 7 millimetres to 4 metres. Maps are commonly drawn with an angular resolution better than 1 arcsecond, comparable with the resolution of many large optical telescopes. At the largest spacing and the shortest wavelength, the angular resolution reaches 4 milliarcseconds, which far exceeds even the angular resolution of the Hubble Space Telescope.

The VLA was a major step in the development of radio telescopes, and it has been outstandingly successful, with literally thousands of papers depending on its observations. It was designed with two new principles. First it was recognized that it was not essential to fill the synthesized aperture with a complete set of interferometer spacings; spacing the separate elements over a larger aperture produced higher angular resolution without a serious penalty in image quality. Second, the array departed from the east–west configuration of the One-Mile Telescope and the Westerbork WSRT, and extended over a two-dimensional pattern. The Y configuration provided at any one time a sufficient sample of interferometer spacings to allow a map to be drawn after an observation lasting only a few minutes. This is the 'snapshot' mode of operation, which has allowed maps to be drawn of a large number of radio sources in a reasonable time. The survey of quasars and radio galaxies which looked for possible lensing (Chapter 5) was made possible by this technique.

The VLA was expensive. It was the first radio telescope with a significant budget on a national scale, having cost US$79 million, approximately 1 dollar for each taxpayer in the USA. Other large projects have followed, but most have been international. The most ambitious so far is the Atacama Large Millimetre Array (ALMA), due to be completed in 2013 (see Chapter 11). Using a total of 66 steerable parabolic reflectors, 12 and 7 metre diameter, sited in Chile at an elevation of 5000 metres, the cost is estimated at around 1 billion dollars. The partners in this major enterprise are North America, Europe, and East Asia.

There are a number of less ambitious synthesis arrays already in operation. At Jodrell Bank Observatory in the UK there has long been a tradition of interferometry using distant antennas linked to the Lovell Telescope by radio. A system which was designed to measure angular diameters, and which had discovered the quasi-stellar radio sources, became the starting point for an aperture synthesis scheme with out-stations, the furthest of which, at Cambridge, was at a distance of 220 kilometres. This became the Multi Element Radio Linked Interferometer, or MERLIN. This is a very sparse array, with only six antennas and large gaps between the available interferometer spacings. Despite this, and with the help of increasingly sophisticated computer techniques, MERLIN has produced some remarkable pictures, most notably of the lensed quasars described in Chapter 5. In 2011 the radio links were superseded by fibre optics, and the improved system became known as eMERLIN. The fibre optic connections can carry a very wide bandwidth, and it is now possible to use a bandwidth of 500 MHz for observations centred on 1500 MHz. This greatly improves sensitivity for any source which covers such a wide band, as indeed do most sources (apart from molecules emitting spectral lines). It also has a remarkable effect on a sparsely populated interferometer network like eMERLIN; since the interferometer baselines are measured as a ratio of wavelength to distance, each antenna pair now covers a wide range, and effectively all gaps in the network are filled. This technique is referred to as 'frequency diversity'.

These large radio telescope arrays were increasingly seen as national or international facilities, using principles which were established in step with technical advances in low noise receivers, broadband fibre optic links, and massive computer power. One further step was an extension of the interferometer baselines to intercontinental, and eventually extraterrestrial, distances.

Very Long Baselines

Connecting radio telescopes together in pairs is a difficult technical task, especially if they are hundreds or even thousands of kilometres apart. The problem is to add together two separate radio signals while preserving their full characteristics. Any oscillating signal, such as a sound wave, is described by its amplitude and phase. Transmitting the two signals to a common receiver requires both amplitude and phase to be preserved; the phase is especially important because the peaks and troughs of the oscillation must add correctly. Adding the signals from different sections of a reflector happens automatically, because they all travel the same distance to the receiver. The signals from two separate telescopes must travel by some different route, which might be a radio link or a fibre optic cable, and yet they must arrive together without distortion.

The first long baseline interferometers, with spacings of around 100 kilometres, used radio links to send the signals to a common receiver. When it became clear that much larger spacings were needed, spanning intercontinental distances, a new technique was developed. At each antenna the signals were recorded on video tape recorders, which used wide and bulky magnetic tape. These were shipped to a central station for playback into a common receiver. A critical requirement was to synchronize the two tapes, so that the signals added correctly in phase. This required a very accurate clock, both at recording and at playback. Despite the complication, and despite the logistical effort of transporting many large cans of tape, this system of Very Long Baseline Interferometry (VLBI) was very successful, providing some of the most detailed maps of distant quasars.

The next stage, using even longer baselines, was to use a radio antenna in a spacecraft as a partner to a ground-based radio telescope. There are immense difficulties in space-based VLBI. Only a small antenna can be mounted in a spacecraft, so that only the most powerful radio sources can be observed. Preserving the correct time relation between the space- and ground-based signals is extremely difficult, because the baseline is changing so rapidly. The achievement of a VLBI Space Observatory Programme (VSOP) by Japan is indeed spectacular. Their satellite, named HALCA, which operated from 1997 to 2005, carried an 8-metre-diameter radio telescope, which opened out like an umbrella when the spacecraft reached its orbit (Plate 16). The baseline between HALCA and Earth-based telescopes reached 21,000 kilometres, three times any possible baseline between pairs of Earth-based telescopes. Maps of quasars were made with angular resolution better than a milliarcsecond. An even better angular resolution, reaching almost to a microarcsecond, is achieved by the Russian Radioastron programme, using a 10-metre-diameter telescope in the satellite Spectr-R, launched in July 2011. This is in a highly elliptical orbit, giving baselines up to 200,000 kilometres, using wavelengths of 1.3, 6, and 92 centimetres. At the shortest wavelength, the angular resolution is measured in microarcseconds. Radioastron opens new possibilities for depicting events very close to the central black holes of quasars. Such angular resolutions are way beyond the capabilities of the best optical telescopes, such as the Hubble Space Telescope, which reach only to around 100 milliarcseconds.

Although space-based VLBI has achieved some spectacular successes, most observations at high angular resolution are made using the less demanding ground-based systems. A dramatic improvement in technique has been the introduction of fibre optic links between pairs of telescopes, at first locally as in eMERLIN but now extending to intercontinental baselines. Optical fibres are very efficient, carrying broadband signals with a stability at least equal to that of radio links; furthermore, they eliminate any need for recording signals and transporting bulky video tapes between observatories. The new generation of radio telescopes, described in Chapter 11, comprise large arrays of individual radio telescopes linked to a central receiver station by optical fibres.

Angles and Distances

Astrometry, the measurement of positions of celestial bodies, is a vital part of radio astronomy, as it has always been in traditional astronomy. Measuring and cataloguing positions requires a frame of reference, like latitude and longitude for terrestrial mapmaking. The difficulty is that there are no fixed points in the sky: the rotation axis of the Earth moves, and stars move. The most distant objects have the most stable positions, and the best of these for astrometry are the quasars, which are prime subjects for VLBI. A grid of quasar positions, measured by radio and covering the whole sky is therefore used as a reference for the most accurate astrometry.

Measuring the angle between the positions of two adjacent objects is always easier than finding a position in relation to a fixed frame of reference. We often need to observe how an object is moving across the sky, and this is achieved by measuring movement in relation to an object such as a quasar which is so far away that its own apparent movement is negligible.

The largest ground-based interferometer array is VLBA, the Very Long Baseline Array in the USA, which comprises 10 identical radio telescopes spanning baselines up to 8000 kilometres, and provides images with resolution of less than a thousandth of an arcsecond. This allows exquisite precision in astrometric measurements. For instance, Shami Chatterjee and colleagues at the National Radio Astronomy Observatory used the VLBA to observe the angular movements of a pulsar PSR 1508+55. This pulsar is one of the fastest moving pulsars, but it is so far away that the angular movements are very small. The velocity was found to be over 1000 kilometres per second, measured with 10% accuracy. But the main achievement, which required observations spread over 2 years, was to measure the annual cyclic change of position due to the earth's orbital motion. This motion, known as parallax, amounts to less than ½ millisecond of arc. Distances of astronomical objects are most accurately measured by their parallax, and distances are often quoted simply as the inverse of the parallax; a parallax of ½ milliarcsecond means a distance of 2 kiloparsecs (kpc), which is 6520 light-years. Conventional methods, used in optical astronomy for centuries, scarcely reach to parallaxes as small as a few milliarcseconds, or distances greater than 1 kiloparsec. The radio measurement has dramatically extended the range of accurate distance measurements, giving the distance of the pulsar as 7700 light-years within an accuracy of 10%.

11

LOFAR, ALMA, and the SKA

Three really large radio telescopes are coming into operation during the decades 2010–2030. Between them they cover the wavelength range from several metres down to less than a millimetre, which is the whole of the range available for ground-based radio telescopes. At even longer wavelengths the terrestrial ionosphere is an insuperable barrier, while at the shortest radio wavelengths the atmosphere is opaque due to absorption by water vapour and other gases. The wide range of wavelengths between these limits is reflected in a wide range of techniques, but the main principles are the same. They all use aperture synthesis.

As we saw in Chapter 10, aperture synthesis builds on the property of radio waves that the signals picked up by pairs of telescopes, which may be separated by large distances, can be combined without loss of their essential characteristics (in technical terms, there is no loss of *coherence*, and both amplitude and phase are preserved). A radio telescope array may therefore comprise a number of individual telescope elements distributed over a large area, acting as a single unit. The angular resolution of the array then depends on its overall extent, rather than the size of its individual elements. For each of the three synthesis telescope arrays described in this chapter there will be two distinct beamwidths to be specified: one is the primary beamwidth, which is the area of sky covered by the individual elements, the other is the much smaller synthesized beamwidth resulting from the combination of the many elements of the array. The way in which radio signals picked up by the individual elements are combined determines the position of the synthesized beam (or multiple beams) within the primary beam.

As an example, in the Very Large Array (VLA), which was described in Chapter 10, the diameter of the individual radio telescopes is 25 metres, while at its maximum the array of 27 telescopes extends over 36 kilometres. At 6 centimetre wavelength each telescope is picking up radio waves over a primary beam 10 minutes of arc across, while their combination in the full array makes a synthesized beam only one third of an arcsecond across. Further more, this narrow beam can be placed anywhere within the primary beam, and with a multiplicity of receiver systems any number of the possible synthesized beams can be in use simultaneously. In terms of conventional photography, the VLA could provide a picture of a large area of the Moon with over 3 million pixels.

Combining the signals picked up by many separate telescope elements has the advantage that the combined collecting area is far larger than is possible in a single radio dish. At the same time, combining the signals from widely spaced telescope elements creates a narrow beam, which is used for mapping finer detail. The technical challenge in this process of *aperture synthesis* lies in the interconnection of the signals from the separate elements, which must be transmitted to a central receiver system without loss and combined with a precise regard to their arrival times. The timing requirement is vital. Figure 58 shows the problem for a single pair of telescopes.

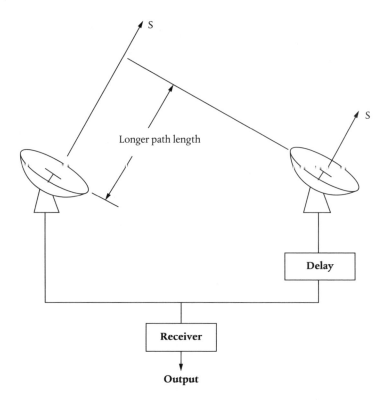

FIGURE 58 Combining radio signals picked up by two separate radio telescopes. The two signal paths must be made equal by inserting a delay into the shorter path; the delay must be adjusted continuously as the source moves across the sky.

The radio signal from any source in the beam arrives earlier in one of the pair, and it must be delayed before the two signals are combined. The delay must be varied according to the position of the source, and the delay must be applied to each pair of signals to be combined in the array. These multiple and very precise delays are achieved by transforming each signal into a digital form, and introducing a delay into the computer which performs the combination. Precise timing is at the heart of any synthesis system; any process which occurs at the individual telescope, such as digitizing, or in the signal transmission, such as adapting the signals for transmission in a fibre optics cable, must be done without loss of timing accuracy.

The design of computer systems which can achieve this precision, and process the combined signals to produce the multiple outputs of the synthesised beams, is very demanding. Very high speeds are required; the signals may be digitized at a rate of several gigahertz, i.e. more than 10^9 operations per second, and many computations are required in parallel for each telescope. Furthermore, it may be possible to operate a large synthesis array at several frequencies simultaneously, looking for example at more than one molecular species in a single gaseous nebula. The computer requirements are a major limitation on the possible uses of synthesis telescopes, and it is usual for designers to assume that progress in computer systems during the several years involved in telescope construction will meet their growing demands.

LOFAR: the Low Frequency Array Radio Telescope

'Low frequency' for radio astronomy means below 300 MHz (wavelengths longer than 1 metre). Most of the early discoveries in radio astronomy were made at low frequencies, where the radio sky is bright and receiver techniques were available. Long wavelengths meant, however, that angular resolution was poor, as can be seen by the efforts required even to measure the size of the brightest radio sources (see Chapter 5). At the lowest frequencies, below about 10 MHz, the terrestrial ionosphere is an opaque barrier which can only be overcome by spacecraft at high altitudes,[69] and even at higher frequencies the ionosphere presents a problem to accurate mapping and position-finding, as it bends radio waves by a variable amount depending on time of day and sunspot activity.

From around 1960 onwards higher frequencies (shorter wavelengths) were used, leaving the low frequencies as a relatively unexplored region of the spectrum. Radio and television broadcasts increasingly occupied much of the available frequency space, and the prospects for low-frequency radio astronomy seemed remote. LOFAR has transformed this situation, and opened a new era in observations.

The difficulty of obtaining angular resolution at long wavelengths is illustrated by a survey by Jelena Milanov-Turin and myself which used the 250-foot (76-metre) Lovell radio telescope in 1967 at a wavelength of 8 metres (38 MHz).[70] Although this telescope was the largest steerable reflector at that time, the beamwidth was 7.5°, 15 times the diameter of the Moon. The synthesis arrays developed at Cambridge (Chapter 10) showed the way out of the problem, extending apertures to a kilometre or more and improving resolution to better than 1°. Moving on to large enough apertures to provide angular resolutions better than 1 minute of arc rather than 1° needed a large-scale development.

LOFAR, built by ASTRON, the Astronomical Institute of the Netherlands, became possible with the advent of two new technologies, fibre optics and massive computer power. The principle is to use a large number of small, fixed, omnidirectional antennas, spread over an area hundreds or even thousands of kilometres across, connected by fibre optic cables to a central computer with huge processing power. No moving parts are involved, but the resulting synthesized telescope beam can be steered over practically the whole sky above the horizon. In the terminology of this chapter, the primary beam covers the whole sky, and the computer can produce simultaneously as many beams as its capacity allows, the width of each being determined by the overall extent of the array. The concept has been described as a 'software telescope', as it depends on a huge effort in programming the combination of signals from thousands of individual antenna elements.

The wide band of frequencies, from 10 to 240 MHz, covered by LOFAR, cannot be covered by a single design of antenna. Abandoning the FM band from 80 to 110 MHz, which is too heavily used for

FIGURE 59 (a) The low band dipoles of the LOFAR array. The pairs of dipoles are centered on the apex of the pyramid, and droop down towards the ground. (b) The high band dipoles are encased in flat tiles, each 5 metres across and containing 16 pairs. © *ASTRON*

broadcasting for any astronomy to be possible, there are separate designs for low (10–80 MHz) and high (110–210 MHz) bands. In each the individual antennas are crossed pairs of wire dipoles, as in Figure 59; in the high band they are enclosed in flat tiles close to the ground, as in Figure 59b. They are combined into clusters, each typically containing some hundreds of dipole units. Each cluster has its own system of interconnections, so that it can operate as an elementary radio telescope with a beam that can be directed in any desired direction. Thirty-six such clusters are in stations spread over an area 100 kilometres in diameter in the Netherlands, and already in 2013 a further eight are located in other European countries, extending the area to 1500 kilometres across. A central core concentrated in 2 × 3 kilometres, is located in the northern province of Drenthe; ASTRON says that the heart of LOFAR beats close to the village of Exloo. Figure 60 shows the disposition of the clusters across Europe.

FIGURE 60 The stations of LOFAR spread across the Netherlands and other European countries. © *ASTRON.*

The fibre-optic links from each station carry signals from the whole wide frequency band, and potentially from several primary beams simultaneously, so that they must operate at a very high data rate, more than a gigabyte per second (my home broadband works more than 1000 times more slowly). The data processing at the central station needs one of the largest computers available, working at tens of teraflops per second.

Although the individual antenna elements are relatively cheap and can be mass produced very economically, the whole system represents a huge investment in design and construction effort as well as in actual cost. The potential for new astronomy is also huge. Out of a wide range of possible observations an outstanding field is cosmology. Hydrogen line radiation from the distant expanding universe originates at a frequency of 1420 MHz (wavelength 21 centimetres), but LOFAR will be able to detect line radiation which has been redshifted by a large factor, reaching us in the low frequency band. The Cosmic Microwave Background, described in Chapter 8, originates in the very early Universe, when ionized gas recombines to form neutral hydrogen and helium. The redshift factor z is around 1000 at this stage. Much later, around redshift $z = 10$. the gas starts to condense into the galaxies and clusters of galaxies which form the present day Universe. Energetic radiation from newly formed bright stars ionises hydrogen again, replacing the all-pervading neutral hydrogen with the ionized clouds familiar in our Galaxy. But the process, and even the timing, of this *epoch of reionization*, is still far from understood. Observing the hydrogen line from the remaining neutral hydrogen in the Universe as it was at redshift $z = 10$, or at a later epoch, may show us this actually in progress. The frequencies involved are simply calculated; they are shifted by a factor $(z + 1)$, putting the redshifted hydrogen line into the upper part of the LOFAR band.

The idea of connecting observing stations spread over a large area of the Netherlands has had an unexpected bonus. Several of the stations have been equipped with entirely different sensing equipment, designed for research in geophysics and climatology. These observations take advantage of the wide-band communication links to investigate, for example the distribution of earth tremors and local weather patterns of interest in agriculture.

LOFAR is proving to be an exemplary international project. As well as European countries with a tradition in astronomical research, others can join in at the cost only of providing another station, which can usefully extend the overall size and sensitivity of the array. In 2012 LOFAR was already operating with stations in Germany, France, Sweden, and the UK.

ALMA: the Large Millimetre Wave Telescope

At the other end of the wavelength range, exploring radio waves with wavelengths 1000 times shorter than LOFAR, millimetre wave astronomy covers a very wide field in astronomy, again including redshifted spectral lines from galaxy formation in the early Universe. The techniques of aperture synthesis are very demanding at such short wavelengths, but they are being applied on an ambitious and impressive scale in the Atacama Large Millimetre Array (ALMA), which was completed in 2013.

The wavelength range of ALMA is 0.3–9.6 millimetres, covering the shortest wavelengths that penetrate the Earth's atmosphere. A very dry site is essential to avoid absorption by atmospheric water vapour, so it has been built at an elevation of 5000 metres in the Atacama desert in Chile. Sixty-six parabolic antennas have been used, with spacings up to 16 kilometres. Every component of this array, from the mounting and surface of each antenna, through the fibre optic connections to the receiver systems has to be accurate to a fraction of the wavelength in use; accuracies are quoted in tens of microns. At this scale of complexity it is not surprising that the cost is over 1 billion dollars.

The millimetre and submillimetre radio wavebands are particularly characterized by a plethora of spectral lines. These originate in molecules, particularly the familiar atmospheric molecules of water, carbon dioxide, oxygen, and nitrogen. Atmospheric spectral lines are a nuisance to astronomers. They both absorb and radiate, obscuring cosmic sources which may themselves be radiating interesting spectral lines.

The multitude of spectral lines shown in the millimetre wavelength range are all due to resonances in molecules, and particularly in their quantized energy levels related to their rates of spin. Provided the effects of atmospheric absorption and radiation are not too large, they can be recognized and allowed for in any analysis of radiation from the sky, with one vitally important exception. Most molecular lines have a constant, or nearly constant effect, but water vapour is distributed in the atmosphere in a very irregular and variable way. Furthermore, there is enough water vapour to have an appreciable effect on the arrival times of radio waves traversing the atmosphere, and the variable delay in arrival time can be very different between the separate units of a synthesis radio telescope. Without corrections, which must involve a calibration process which can be reviewed at frequent intervals, no synthesis array can operate at the millimetre and submillimetre wavebands. Fortunately this has been shown to be possible.

Correction for these atmospheric effects is achieved by deliberately observing in prominent absorption bands, particularly those due to water vapour at 22.2 GHz and due to oxygen at 60 GHz and 118 GHz. The large effects on signals from a prominent compact source, usually a quasar, can then be used to make smaller corrections to the regular observations at frequencies outside these major resonances. Another technique, which is more spectacular to the casual sightseer, is to switch observations alternately between the source to be mapped and a known compact source in a nearby area of sky. The corrections needed can again be found from the effect on signals from the known calibration source. Atmospheric conditions can change rapidly, so that the pointing direction of all antennas must be switched between targets within a few seconds, and repeated within a minute or two. The whole array may be involved simultaneously in this twitching movement.

The dry desert site of the ALMA array (Plate 17) has the advantage that little other use can be made of the site, so that there is practically no radio interference and no competition from agriculture. The disadvantage is the impossibility of pursuing a normal life at an altitude of 5000 metres; those who do work at the site must live in pressurized and air-conditioned laboratories. Most of the construction work on the 66 antennas has to be done at a lower level headquarters, at 2900 metres elevation, where each antenna is assembled and tested. It must then be transported by road up to the telescope site. That is not the final move, since at the site it is intended that the array can be reorganized between observing sessions, to give a more or less compact configuration. Each antenna weighs 115 tonnes, so that even moving one is a major engineering task. Two specially designed transporters are used, each with 28 wheels.

The design of ALMA allows for observations in wide or narrow frequency bandwidths over the whole of the millimetre wavelength band. With full and meticulous correction for atmospheric effects, it should provide images with angular resolution approaching 0.1 arcseconds, and with a high dynamic range

The successful achievement of millimetre wave synthesis, using arrays of telescopes at high altitude and with all the technical challenges of maintaining coherence at such short wavelengths, has already been demonstrated in two telescope arrays, the SubMillimeter Array (SMA) at 4080 metre altitude on Mauna Kea, and at the IRAM Plateau de Bure (PdeB) interferometer at 2550 metres in the French

Alps. Both are already operating in the same range of wavelengths as ALMA. SMA has eight steerable parabolic antennas, each 6 metre diameter; PdeB has six antennas each 15 metres in diameter.

The PdeB interferometer has demonstrated the feasibility of one of the design aims of ALMA, which is to observe molecular spectral lines from distant galaxies with high redshifts. A large-scale survey using UKIRT, the 3.8-metre infrared telescope on Mauna Kea, located one of the most distant galaxies ever seen, at redshift $z = 7.1$. This galaxy, J1120+0641, contains a black hole with the mass of 3 million Suns; it is particularly interesting because it gives us a rare glimpse of the early Universe when galaxies had only recently been formed. There is a large gap in our knowledge of the Universe between the age of recombination at 400,000 years, observed as the Cosmic Microwave Background, and around one billion years, when most galaxies and stars formed. J1120+0641 is observed as it was at age 740 million years, about 13 billion years ago. It has a high star formation rate and a high carbon content, which is surprising in a galaxy formed so early in the evolution of the Universe. The PdeB interferometer observed a spectral line from the interstellar carbon atoms, generated at the short wavelength of 158 microns but redshifted to 1.276 millimetres (frequency 235 GHz), within an atmospheric transparent window.[71]

SKA, the Square Kilometre Array

The culmination of all the experience, aspirations, and hopes of the many radio astronomers who have opened a new view of the Universe through the progressive development of more complex and more sensitive radio telescopes is the SKA, Square Kilometre Array. The crucial steps in this development have been from the simple interferometer to Ryle's One-Mile Synthesis Telescope, and then to the VLA, whose observations figure in thousands of scientific papers every year. The next step is to build a telescope with sensitivity, angular resolution, and spectral coverage all improved by at least an order of magnitude. Some of the new observations that would be possible with such a telescope can already be predicted; it is however such a huge step forward that it will open new, unforeseen fields.

The SKA will cover a large part of the radio astronomy spectrum, between the low frequencies of LOFAR and the millimetre wavelengths of ALMA. It is therefore primarily centred on centimetric wavelengths, between 1 centimetre and 1 metre. Every branch of astronomy will be addressed in this range, and each has proposed its own programmes of research, from the solar system to the history of the early Universe. The key scientific projects which will depend on the sensitivity of SKA include many relating to the early stages of galaxy formation, for example an investigation of the origin of magnetism in galaxies, particularly our own, through measurements of polarization. The formation of molecules, from the simplest such as carbon monoxide to the long carbon chains that approach the complexity of basic organic living molecules, will be part of a major project with the grand title of the Cradle of Life.

Two specific examples will demonstrate the capability of the VLA to open up new fields in astronomy. These are fields in which I have long had a personal interest, pulsars and radio galaxies.

Most simply, the increased collecting area the SKA gives correspondingly greater sensitivity to small distant radio sources, such as the pulsars. The 2000 pulsars so far known include some extremely interesting objects, as outlined in Chapter 7. Ten times more will certainly include many more millisecond pulsars with accurate timing, which is confidently predicted to lead to the detection of large-scale gravitational waves. We have already seen that millisecond pulsars in binary systems are exploring relativity theory to new depths; what is now needed is examples of binaries in which a pulsar is moving in orbit round a black hole, which will test the behaviour of massive bodies in very strong gravitational fields.

On the largest scale, our knowledge of the Universe is weakest at the age of galaxy formation. The millimetre wavelengths of ALMA explore this age through redshifted spectral lines, such as that from carbon, observed at 1 millimetre but emitted at a wavelength six or seven times shorter when the Universe was younger and more compact. SKA will observe redshifted radio waves from every type of emission from the newly formed galaxies of this era; for this it is essential to make maps in very fine detail, to distinguish the redshifted sources from the myriad of foreground objects. To have sufficient angular resolution, the large collecting area of SKA must be spread out over distances extending to thousands of kilometres.

The key scientific projects which will depend on this new capability include an investigation of the origin of magnetism in galaxies, particularly our own, through measurements of polarization. The formation of molecules, from the simplest such as carbon monoxide to the long carbon chains that approach the complexity of basic organic living molecules, will be part of the Cradle of Life project. These examples are only an indication of the capability of SKA.[72]

The SKA will therefore have some hundreds of separate telescope units, which must be connected together with broadband fibre optic links. The total data flow in the complete system will be comparable to the entire present-day digital communication traffic in the whole world, while the computer systems which correlate the signals and produce maps will be at the cutting edge of technology. The choice of number and type of the antenna units, their disposition, their interconnection, and the organization of their correlation in the central computer, have been the subject of many design studies for more than a decade.

A further vital question has been where such a huge telescope should be built. It has to be far from intensive civilization, so as to reduce the effect of man-made radio signals. It should preferably be in the Southern Hemisphere, so that the centre of our Milky Way galaxy will be seen directly overhead. Any country with a suitable site must act as a host for an international project on an almost unprecedented scale. Two sites, in Western Australia and in South Africa, were investigated in great detail, leading to a decision in 2012, Both were found to be suitable, and eventually it was decided to divide the project between the two sites.

For several years before the decision on splitting the SKA between the two sites, the new technologies needed for the SKA were being demonstrated in several precursor and pathfinder telescopes, particularly in the two chosen locations. In South Africa a large area of the Karoo, a semi-desert area in Northern Cape Province, has been designated for the SKA, and an array of 64 steerable parabolic reflectors known as MeerKAT is under construction at this site. An array of seven such dishes, KAT-7, is already in operation, and an extensive infrastructure of laboratories and power supplies is in place. Most importantly, institutes and universities throughout Africa are extending their educational resources towards developing the new technologies and research manpower which will be needed for SKA.

The Australian bid is centred on the Murchison Radio Observatory in the outback of Western Australia. Here there is already a low-frequency array, the Murchison Widefield Array, consisting of 128 'tiles' each with 16 crossed dipoles, for frequencies 80–300 MHz. There is also a pathfinder array for the shorter wavelength component of the SKA known as ASKAP, for which 36 steerable parabolic antennas are already in place.

The decision to split the SKA allocates the main array of several thousand parabolic dishes to South Africa, and the low frequency array of 'tiles' to Australia. This augments existing pathfinder arrays in both locations, and uses established infrastructure, but it does of course involve duplication of the fibre optic network and part of the computer correlator. There is an overwhelming political and social advantage in developing both sites, since the network of antennas will spread into more neighbouring countries; Tasmania and New Zealand will be part of the Australian network, while the African project will involve Madagascar and Mauritius as well as several adjacent countries on the African continent. All these countries will benefit by access to front-line astronomical research and by technological advances, which will inspire a young generation of scientists and engineers.

The statistics of the SKA itself are hard to grasp. Three different antenna types will be used, to cover the very wide frequency range from 70 MHz to 10 GHz. The high frequency antennas will be 3000 fully steerable parabolic reflectors, while the lower frequencies will use assemblies of dipoles like those of LOFAR, the beam of each being steerable electronically by phasing the connections between dipoles. Both types will be densely distributed in the central region of the SKA, with about half spread progressively further apart, ideally along five spiralling radial arms. The outermost parts of these arms will spill over into adjacent countries, reaching a distance of 3000 kilometres. All antennas must be connected by fibre optic cable, making a total of some thousands of kilometres of cable. Each cable will carry multiple frequency channels, allowing fine resolution of spectral lines and rejection of interference by narrow-band man-made radio signals. The central computer must perform 10^{17} operations per second: that is off the scale of common engineering parlance, where we have become used to the gigabit (10^9) and the terabit (10^{12}) and even the petabit (10^{15}). The capacity of this central computer will be greater than the current global internet traffic. Even the output, in which the input data have been boiled down to only the wanted spectra and maps, will be at the rate of 10 gigabits per second.

To achieve all this at a minimum cost, radical designs are needed for every component. The reflector telescopes will be 15 metres diameter;

FIGURE 61 The central concentration of parabolic reflector antennas of the Square Kilometre Array. *SKA Organisation/Swinburne Astronomy Productions.*

an array of these in the close configuration of the central array, is seen in Figure 61. The energy requirements of the whole system could be met by several conventional generating stations, but will largely be met by innovative technologies in local generation and storage.

The SKA will be built in two phases, over a period of about 8 years. Observations are planned to start in 2019, with the performance progressively improving through the second phase up to completion in 2025 (Figure 62). Financing and managing this huge enterprise is spread over many countries, and all will take part in the choice of observing programmes.

The Office of the SKA Organization is located at the Jodrell Bank Observatory in Cheshire, UK.

The current partners in the SKA project (in 2013) are

- Australia: Department of Innovation, Industry, Science, and Research
- Canada: National Research Council

FIGURE 62 The arrangement of the outer parts of the Square Kilometre Array. Curved arms extend from the central array, with more isolated individual dishes extending for hundreds of kilometres. *SKA Organisation/Swinburne Astronomy Productions.*

- China: National Astronomical Observatories, Chinese Academy of Sciences
- Italy: National Institute for Astrophysics
- New Zealand: Ministry of Economic Development
- Republic of South Africa: National Research Foundation
- The Netherlands: Netherlands Organisation for Scientific Research
- United Kingdom: Science and Technology Facilities Council.

More than 70 institutes in 20 countries are involved in the design of the SKA.

It has been a long journey from the first radio telescopes and interferometers described in the early chapters of this book. The whole story has been part of my own life, and now I can only stand back and wonder what astronomy will be like after another decade or two.

Notes

Chapter 1

1. Radio waves are usually specified by their frequency. The 15-metre wavelength used by Jansky has a frequency of 20 megahertz (MHz). Most broadcast radio receivers now work in the VHF band, at a frequency of around 100 MHz, with a wavelength around 3 metres. The product of wavelength and frequency for all electromagnetic waves in a vacuum is the speed of light. Hence the product of wavelength in metres and frequency in MHz is 300.
2. Quoted by Woodruff T. Sullivan III, in *Cosmic Noise*, Cambridge University Press, 2009, which expounds the early history of radio astronomy in fascinating detail.
3. This map of the radio sky was published by Haslam C.G.T., Salter C.J., Stoffel H., Wilson W.E. 1982. A 408 MHz all-sky continuum survey. II—The atlas of contour maps. *Astronomy and Astrophysics* **47** (suppl), 1.

Chapter 2

4. Sir Martin Ryle (knighted 1966) was an outstanding pioneer of techniques for radio astronomy. He was the first professor of radio astronomy at the University of Cambridge, and Astronomer Royal 1972–82. In 1974 Ryle and Antony Hewish shared the first Nobel Prize in Physics awarded in the field of astronomy.
5. Antony Hewish FRS (b. 1924). British radio astronomer who won the 1974 Nobel Prize for Physics (jointly with fellow radio-astronomer Martin Ryle) for his work on the development of radio aperture synthesis and for the discovery of pulsars.
6. James Hey FRS (1909–2000) is credited with not just one but three of the major discoveries which started radio astronomy after World War II.

Apart from the Sun, he was the first to find radar echoes from meteor trails, a story followed up by Bernard Lovell at Jodrell Bank, and he was the first to find the object which we now know as the distant radio galaxy Cygnus A.

7. Dr John Paul Wild AC CBE FRS (1923–2008) was a British-born Australian scientist who served in World War II as a radar officer in the Royal Navy. He led the solar radio research team at the Commonwealth Scientific and Industrial Research Organisation in a very productive period in the 1950s and 1960s, and eventually became Chairman of CSIRO. The Paul Wild Observatory at Culgoora in New South Wales is named after him.

8. See a review by I. de Pater 1990 Radio images of the planets. *Annual Review of Astronomy and Astrophysics* **28**, 347.

9. Prof. Bernard F. Burke, William A.M. Burden Professor of Astrophysics, formerly of the Radio Astronomy Group of the Research Laboratory of Electronics, is now a principal investigator at the MIT Kavli Institute for Astrophysics and Space Research.

10. Burke B.F., Franklin K.L. 1955. Observations of a variable radio source associated with the planet Jupiter. *Journal of Geophysical Research* **60**, 213.

11. Hugill J., Smith F.G. 1965. Cosmic radio noise measurements from satellite Ariel II. I, Receiving system and preliminary results. *Monthly Notices of the Royal Astronomical Society* **131**, 13.

12. Moon radar: it appeared some time after the war that the German long range radar Wassermann had seen moon echoes in 1942, as reported by David Pritchard in *The Radar War*, Patrick Stephens 1989.

13. Zoltan Bay's heroic efforts are detailed in Sullivan, *Cosmic Noise*, Cambridge University Press 2009 pp. 271–4.

14. For a comprehensive review of planetary radar up to 1965 see Pettengill G.H., Shapiro I.I. 1965. Radar astronomy. *Annual Review of Astronomy and Astrophysics* **3**, 377.

15. Pettengill G.H., Dyce R.B. 1965. A radar determination of the rotation of the planet Mercury. *Nature* **206**, 1240.

16. The first radar tests of the Shapiro delay were reported in Shapiro I.I., Pettengill G.H., Ash M.E., Stone M.L., Smith W.B., Ingalls R.P., Brockelman R.A. 1968). Fourth test of general relativity: preliminary results. *Physical Review Letters* **20** (22), 1265–9.

Chapter 3

17. Oort J.H., Kerr F.J., Westerhout, G. 1958. The galactic system as a spiral nebula. *Monthly Notices of the Royal Astronomical Society* **118**, 379.
18. Astronomers usually specify these large distances in kiloparsecs (kpc). 1 kpc = 3260 light-years.
19. Brinks E. & Shane W.W 1984. Astron.& Astrophys. Suppl. 55, 179. A high resolution hydrogen line survey of M 31.
20. Yakov Borisovich Zel'dovich (1914–1987) was a prolific Soviet physicist who made important contributions to astrophysics, cosmology, and general relativity.
21. Balick B., Brown R.L. 1974. Intense sub-arcsecond structure in the galactic center. *Astrophysical Journal*, **194**, 265.
22. Lynden-Bell D. 1969. *Nature* 223, 690.
23. 300,000 km s^{-1} is a convenient approximation for the velocity of light; the actual definition is 299,792,458 metres per second.
24. The modern unit of magnetic field is the tesla, named in 1960 after Nikola Tesla (1856–1943), physicist, electrical engineer and inventor; 1 tesla equals 10^4 gauss.
25. Smith F.G. 1950. Origin of the fluctuations in the intensity of radio waves from galactic sources: Cambridge observations. *Nature* **165**, 422.
26. Machin K.E., Smith F.G. 1952. Occultation of a Radio Star by the Solar Corona. *Nature* **170**, 319.

Chapter 4

27. Patrick Blackett. Baron Blackett OM CH FRS was an important influence on the development of radio astronomy after World War II. He served in the Royal Navy in WWI, and advised on military strategy in WWII. In 1948 he was awarded a Nobel Prize for his investigations of cosmic rays.
28. Galbraith W., Jelley J.V. 1953. Light pulses from the night sky associated with cosmic rays. *Nature* **171**, 349.
29. Smith F.G., Porter N.A., Jelley J.V. 1965. *The detection of radio pulses of wavelength 6.8 m, in coincidence with extensive air showers, in the energy region* 10^{16}–10^{17} *eV*. Proceedings of the 9th International Cosmic Ray Conference, **I**, p. 701.

30. Details of HESS and MAGIC can be found in their excellent websites.
31. Hannes Alfvén is well known for his discovery of a type of wave motion in an ionised gas which is pervaded by a magnetic field. These Alfvén Waves occur in many different situations, including the atmospheres of the Earth and the Sun.
32. Polarization. In any beam of electromagnetic radiation, the electric and magnetic fields are aligned across the direction of the ray, and at a right angle to each other. In a linearly polarized radio wave they are aligned in a fixed direction, anywhere at a right angle to the ray direction. In an unpolarized radio wave the field direction changes randomly. In a circularly polarized wave the field direction rotates uniformly around the ray direction, either clockwise or anticlockwise; this is known as right- or left-hand circular polarization.

Chapter 5

33. John Gatenby Bolton FRS (1922–1993) was a British-Australian astronomer from Sheffield, UK. After graduation from Cambridge he joined the Royal Navy, serving on HMS *Unicorn* during World War II. His ship went to Australia; he remained there after the war, and in 1946 he began working at the CSIRO Division of Radiophysics.
34. Unwin S.C., Cohen M.H., Pearson T.J., Seielstad G.A., Simon R.S., Linfield R.P., Walker R.C. 1983. *Astrophys J* 272, 383; Pearson T.J., Unwin S.C., Cohen M.H., Linfield R.P.; Readhead A.C.S., Seielstad G.A., Simon R.S., Walker R.C. 1981. *Nature* 290, 365.
35. Martin John Rees, Baron Rees of Ludlow, OM, Kt, FRS (b. 23 June 1942) Astronomer Royal (since 1995) and Master of Trinity College, Cambridge from 2004 to 2012. He was President of the Royal Society between 2005 and 2010.
36. Walsh D., Carswell R.F., Weymann R.J. 1979. 0957 + 561 A, B - Twin quasistellar objects or gravitational lens. *Nature* **279**, 381.

Chapter 6

37. Pulsars are catalogued by their positions in the sky, (right ascension and declination, similar to geographical longitude and latitude). The

designation B refers to 1950 coordinates; most positions are designated as J, based on 2000 coordinates.

38. Dispersion is the effect of propagation through the free electrons in interstellar space. The delay is typically only a few seconds, even though the pulses may have been travelling for thousands of years. The delay is greater at low radio frequencies.

39. Lyne A.G. 1977. A new pulsar survey. Proceedings of the Astronomical Society of Australia 3, 118–19.

40. Degeneracy. More precisely, in a gas the fermionic nature of matter means that particles cannot occupy the same state, so that there is a quantum mechanical force which prevents a collapse to infinite density. This force determines the maximum compression the material can sustain, when it becomes degenerate. In white dwarf stars the density is determined by the degeneracy pressure of electrons, and in neutron stars by the degeneracy pressure of neutrons.

41. Hankins T.H., Eilek J.A. 2007. Radio emission signatures in the Crab pulsar. *Astrophysical Journal* 670, 693.

Chapter 7

42. White dwarf stars. Many stars end their evolution when the core of the star collapses to become a white dwarf, with a diameter similar to that of the Earth, but with a density a million times greater.

43. Weisberg J.M., Nice D.J., Taylor J.H. 2010. Timing Measurements of the relativistic binary pulsar PSR B1913+16. *Astrophysical Journal,* **722**, 1030.

44. Shapiro I.I. 1964. Fourth test of general relativity. *Physics Revue Letters* **13**, 789.

45. The two components of the binary pulsar are designated PSR J0737-3039A and B. The discovery of B was published in Lyne A.G., Burgay M., Kramer A., *et al.* 2004. A double-pulsar system: a rare laboratory for relativistic gravity and plasma physics. *Science* **303**, 1153.

46. See a description by Graham-Smith F., McLaughlin M.A. 2005 A magnetopause in the double-pulsar binary system. *Astronomy and Geophysics* **46,** 24.

Chapter 8

47. Magnitude is a logarithmic scale, in which 5 magnitudes corresponds to a factor of 100 in intensity. The log N proportional to $0.6m$ relation is

easily derived from the inverse square law of the decrease of intensity S with distance, combined with the cube law of the increase of volume of space. This gives the relation N proportional to $S^{-3/2}$, which is the same law: it is the way that radio astronomers expressed the number counts during the controversies of the 1950s and 1960s.

48. Hubble's original value for the age of the Universe, based on his value for the Hubble Constant, was 2 billion years, which could not be right since it was less than the known ages of the oldest rocks on Earth. The age is now known to be close to 14 billion years.

49. Alpher R.A., Bethe H., Gamov G. 1948. The origin of the chemical elements. *Physical Review* **73**, 803.

50. Mitton S. 2011. *Fred Hoyle: a life in science.* Cambridge University Press.

51. Windhorst R.A., Miley G.K., Owen F.N., *et al.* 1985. Sub-millijansky 1.4 GHz source counts and multicolor studies of weak radio galaxy populations. *Astrophysical Journal* **289**, 494.

52. Redshift z is defined as the relative change in wavelength $\Delta\lambda/\lambda$. The redshift depends on the ratio of the source velocity v to the velocity of light c. $1+z=\dfrac{c}{c-v}$.

53. The geometrical description of the expanding Universe is the Robertson-Walker metric. The effect on the number counts for sources at large redshifts is described in detail by Malcolm Longair in Section 17.2.2 of *Galaxy Formation*, 2nd edition, 2008.

54. A more precise formulation combines the effect on space and time together, so that we can speak of a curvature of 'space–time'. A straight line in bent space-time then becomes a 'geodesic', which is the minimum path between two events.

55. Alexander Alexandrovich Friedman (1888–1925) was a brilliant meteorologist who applied his understanding of the flow of gases and liquids to Einstein's theory. He published his results in 1922 and 1924, but he died of typhoid in Leningrad before the significance of his work was appreciated. The same results were found independently by the Belgian priest Abbé George Lemaître (1894–1966) in 1927. It is remarkable that both these men were discussing the expansion (and possible contraction) of the Universe several years before Hubble discovered the velocity–distance relationship which is the basis of modern cosmology.

56. The results of the observations establishing the reality of the cosmic acceleration were published in two papers: Riess A., Filippenko A.V., Challis P., *et al.* 1998. Observational Evidence from supernovae for an accelerating universe and a cosmological constant. *Astronomical Journal* **116** (3): 1009; Perlmutter S., Aldering A., Goldhaber R.A., *et al.* 1999. Measurements of omega and lambda from 42 high-redshift supernovae. *Astrophysical Journal* **517** (2), 565.

57. Guth A. II. 1981. The inflationary universe: a possible solution to the horizon and flatness problems. *Physical Review* **D23**, 347.

58. An exception: gravitational waves originating in the inflationary era might eventually be observable.

Chapter 9

59. A black body ideally completely absorbs every type of electromagnetic radiation that falls on it. Following basic laws set out by Planck, radiation from a black body has an intensity and spectrum which depend only its temperature. The peak of the 'blackbody' spectrum, at a wavelength λ_{max} (in metres), is related to temperature T (in degrees K) by Wien's Law $\lambda_{max}T \approx 2.9 \times 10^{-3}$ mK.

60. Mather J.C., Cheng E.S., Eplee R.E. *et al.* 1990. A preliminary measurement of the cosmic microwave background spectrum by the Cosmic Background Explorer (COBE) satellite. *Astrophysical Journal* **354**, 37.

61. Gush H.P., Halpern M., Wishnow E.H. 1990. Rocket measurement of the cosmic-background-radiation millimetre-wave spectrum. *Physics Revue Letters* **65**, 537.

62. Jones W.C. and 36 others. 2006. A measurement of the angular power spectrum of the CMB temperature anisotropy from the 2003 flight of BOOMERANG. *Astrophysical Journal* **647**, 823.

63. Bennett C.L. and 20 co-authors, 2003. First-year Wilkinson Microwave Anisotropy Probe (WMAP) observations: preliminary maps and basic results. *Astrophysical Journal* (Suppl) **148**, 1.

64. Spherical harmonics describe the structure of the ripples in the background temperature across the sky. Structure with a typical angular size θ radians (one radian equals 57.3°) has an *l* number equal to π/θ. The individual patches in the predominant pattern of the CMB have a size of

around 1° (about twice the diameter of the Moon), or 1/57 radian, so the strongest spherical harmonic is at l=180.

Chapter 10

65. See *The Story of Jodrell Bank*, Bernard Lovell, Oxford University Press, London and New York, 1968.
66. The angular resolution of a circular aperture with diameter d at wavelength λ is often quoted as 1.22 λ/d. This is quoted in radians; one radian is 57.3°, or approximately 3400 arcminutes, or 200,000 arcseconds.
67. Joseph Pawsey was the first to remark that the interferometer was responding to a Fourier component of the brightness distribution across the source, a formulation which appeals to mathematicians and engineers.
68. Ryle M., Neville A.C. 1962. A radio survey of the North Polar region with a 4.5 minute of arc pencil-beam system. *Monthly Notices of the Royal Astronomical Society* **125**, 39.

Chapter 11

69. The first satellite-borne radio astronomy observation was in Ariel II, launched in 1964, (Hugill and Smith, F.G. 1965 MNRAS 131, 137). The receiver scanned frequencies from 1 to 3.5 MHz, using a dipole 40 metres long (short compared with a wavelength of 300 metres at 1 MHz).
70. Milogradov J., Smith F.G. 1973. A survey of the radio background at 38 MHz. *Monthly Notices of the Royal Astronomical Society* **161**, 269.
71. J1120+0641, the galaxy with redshift z = 7.1, was discovered at UKIRT by Mortlock D.J., Warren S.J., Bram P.V., *et al.* 2011. A luminous quasar at a redshift of z = 7.085. *Nature* **474**, 616. The millimetre wave spectral line detection was detected at Plateau de Bure by Venemans B.P., McMahon R.G., Walter F., *et al.* 2012. Detection of atomic carbon [C II] 158 μm and dust emission from a z = 7.1 quasar host galaxy. *Astrophysical Journal* **751**, 25.
72. For an exposition of the scientific programmes of the SKA, see Carilli C.L., Rawlings S. 2004. Science with the Square Kilometre Array. *New Astronomy Reviews* **48**, 979.

Index